SEMIGROUPS OF LINEAR AND
NONLINEAR OPERATIONS AND APPLICATIONS

Semigroups of Linear and Nonlinear Operations and Applications

Proceedings of the Curaçao Conference, August 1992

Edited by

GISÈLE RUIZ GOLDSTEIN

and

JEROME A. GOLDSTEIN

Department of Mathematics,
Louisiana State University
Baton Rouge, Louisiana, U.S.A.

SPRINGER SCIENCE+BUSINESS MEDIA, B.V.

A C.I.P. Catalogue record for this book is available from the Library of Congress.

ISBN 978-94-010-4834-7 ISBN 978-94-011-1888-0 (eBook)
DOI 10.1007/978-94-011-1888-0

Printed on acid-free paper

Table of Contents

Director's Preface 3

Editors' Preface 7

Jerome A. Goldstein:
A Survey of Semigroups of Linear
Operators and Applications 9

Gisèle Ruiz Goldstein:
Nonlinear Semigroups and Applications 59

Wim Caspers and Philippe Clément:
A Bifucation Problem for Point Interactions in $L^1(\mathbb{R}^3)$ 99

Alfonso Castro and Ratnasingham Shivaji:
Semipositone Problems 109

Iona Cioranescu:
A Generation Result for C-regularized Semigroups 121

Giuseppe Da Prato:
Smoothing Properties
of Heat Semigroups in Infinite Dimensions 130

W.E. Fitzgibbon, S.L. Hollis, and J.J. Morgan:
Locally Stable Dynamics for Reaction-Diffusion Systems 143

W. Fitzgibbon, M. Parrott, and Y. You:
Global Dynamics of Singularly
Perturbed Hodgkin-Huxley Equations 159

Matthias Hieber:
On Strongly Elliptic Differential Operators on $L^1(\mathbb{R}^n)$ 177

Alessandra Lunardi:
Stability and Local Invariant Manifolds in
Fully Nonlinear Parabolic Equations 185

Adam C. McBride:
Fractional Integrals and Semigroups 205

Rainer Nagel:
Spectral and Asymptotic Properties of
Strongly Continuous Semigroups 225

J. W. Neuberger:
Continuation for Quasiholomorphic Semigroups 241

Michel Pierre and Didier Schmitt:
Global Existence for a Reaction-Diffusion System
with a Balance Law 251

G. F. Webb:
Convexity of the Growth Bound of
C_0-semigroups of Operators 259

Index 271

Director's Preface

This is the first publication which follows an agreement by Kluwer Publishers with the Caribbean Mathematics Foundation (CMF), to publish the proceedings of its mathematical activities. To which one should add a disclaimer of sorts, namely that this volume is not the first in a series, because it is not first, and because neither party to the agreement construes these publications as elements of a series. Like the work of CMF, the arrangement between it and Kluwer Publishers, evolved gradually, empirically.

CMF was created in 1988, and inaugurated with a conference on Ordered Algebraic Structures. Every year since there have been gatherings on a variety of mathematical topics: Locales and Topological Groups in 1989; Positive Operators in 1990; Finite Geometry and Abelian Groups in 1991; Semigroups of Operators last year. It should be stressed, however that in preparing for the first conference, there was no plan which might have augured what came after. One could say that one thing led to another, and one would be right enough.

Yet, that is not quite the whole story, For a long time it had been clear to me that the Caribbean Basin and Latin America languished in a general state of mathematical oblivion. And having said that I'm reminded that one should be careful with such statements, that there certainly are a number of distinguished mathematicians from this part of the world, and also a number of centers where first-rate work is being carried out. That is not the point, however; what is the raison d'etre of CMF is that the men and women and centers of distinction in mathematics in the Caribbean Region and Latin America are few and far between; that the vast majority of mathematicians there gradually lose contact with the developed world because they are isolated in underdevelopment, cut of from physical contact with experts. They are isolated because their libraries are persistently out of date.

CMF enters this stage of underdevelopment with a mission which is easy to formulate and yet difficult to carry out. The mission is this: to do all it can to diminish the isolation alluded

G. R. Goldstein and J. A. Goldstein (eds.),
Semigroups of Linear and Nonlinear Operations and Applications, 3–6.

to in the preceding paragraph. It is a difficult task for two main reasons. First, it is not always clear how one can do the most good. (More on this subject shortly.) And even when it is, there are problems, faced by mathematicians of the region, which are so typical of underdevelopment: poverty, institutional paternalism and bureaucracy. We, in our cozy, developed universe, who complain about the shortage of funds when it comes to travel to conferences, should have some idea of the shortage faced by mathematicians in the so-called Third World. As it seems inappropriate to use these pages to compile woes, let me illustrate with one example: a mathematician from El Salvador, Nicaragua, Surinam or Guyana will typically get no funds at all from his/her institution or home country, even when there is an explicit letter of invitation.

But perhaps one need not strain to document poverty in underdevelopment. Institutional paternalism, however, is less palpable, and the effects of bureaucracy certainly as cruel as those of poverty. The Spanish have a word for the kind of "red tape chase" that takes days, weeks, months, and saps one's enthusiam: trámite. One Colombian colleague recently quipped that his country was the land of "tramitologia".

In the developed world a mathematician who has made plans to attend a conference simply arranges for a colleague to take over his classes for a few days and leaves. Not so in the Caribbean Region or Latin America, in general. There one has to submit a formal request for leave, which then travels through the entire bureaucratic apparatus of a university. If an academic goes to a conference without official permission, he runs a great risk of not being paid for the time away.

Such are the problems, even in the presence of the best laid plans. The good news is that there are enough mathematicians in the developed world with a true missionary spirit to challenge the condition of underdevelopment in this region. It helps to be able to invite one's colleagues to a tropical island like Curaçao, where the swimming and snorkeling are quite satisfactory, and where the Dutch façades on the St. Anna Bay strike the typical visitor with more than a touch of wonder. And yet, not every distinguished mathematician is suited to confront the special problems of underdevelopment. Even the most altruistic soul has seen the reality and been disappointed: at the low attendance by Caribbean and

Latin American mathematicians at the workshops. I have seen the disappointment in the faces of my colleagues, and shared in it.

Which brings me round to the first-mentioned problem, in discharging the simple mission of CMF. Deciding what themes and topics to choose for the workshops and conferences.

Doubtless the attendance at these events by mathematicians from the region is small because of the poverty and bureaucracy to which I have already referred. However, one should also realize that the region in question is huge, and that the interests of its mathematicians varies enormously, not only in terms of discipline or speciality, but also in terms of level of development. Last year's workshop/conference on *Semigroups of Operators and Applications* is a good case in point. The conference attracted four South American mathematicians: two from Chile and two from Colombia. One might stare at that figure and say it is small. My experience tells me that to draw four mathematicians from South America constitutes an enormous success. The workshop portion of the program drew another three participants from Surinam and one Jamaican. What this number does not reveal is that the latter four have become regulars at these events. Two of the Surinamers have been attending since 1990, the third since 1991, the Jamaican since the beginning.

One comes to an attitude of compromise and resignation, while not losing sight of the goal, as stated above. Missionaries of true spirit know how frustrating the task of promulgating the Gospel can be. One has to know that the job is worth doing: it is good to disseminate mathematics in the Caribbean Region and Latin America. In the meantime one tries to find out how best to serve its mathematical community, knowing that there will be moments of frustration along the way; that one will make mistakes along the way. One has to know that the crowds will be small; that the participants who return, year after year, for whatever reason, are, in the long run, the best promotors one has.

One counts one's blessings; the greatest two being to be able to rely on the services of colleagues who instinctively understand that the problem of doing something about underdevelopment is not so simple, yet who by their talent and spirit contribute to the realization of CMF's goals, with excellent programs. No one has done this with more grace and verve than Jerry Goldstein and

Gisele Ruiz Goldstein, through their program on *Semigroups of Operators and Applications*. Generally, I have a solid expectation that the colleagues with whom I contract to prepare a workshop/conference for CMF will acquit themselves well. Indeed, I can be more bold: I count on the people selected to present a program under the auspices of CMF, which will produce an event of the highest quality, and also display the kind of sensitivity one needs when coming face to face with underdevelopment. I have not yet been disappointed on either point.

During the three days of the workshop, the Goldsteins presented an account of both the linear and non-linear theories of semigroups of operators, by clearly grounding the subject both in terms of its origins in physics and of the philosophy of its development, returning with regularity to a number of illustrative examples. Perhaps it is not quite fair to volunteer myself as a judge, but, as decidedly a non-expert in this field, I was able to see the logic from formulation to execution, and understand, if very few of the details, at least the mathematical tools that are brought to bear. It seems to me that one cannot ask more of such a workshop presentation.

As to the conference which followed, I can only judge by the comments of its participants. I spoke to almost all of the twenty or so, and each was thoroughly satisfied with the mathematical event. We plan to reprise a conference on this discipline before the end of the decade.

I spoke earlier of two blessings; I've expatiated on one, but have not forgotten the other. One is fortunate to have in Dr. David Larner and his staff at the Science and Technology Division of Kluwer Academic Publishers people who have understood the problem of disseminating mathematics in the region, and have backed this understanding and their encouragement of CMF with deeds. Their agreement to be the publishers of these proceedings is but one of them.

Jorge Martinez, Director

Caribbean Mathematics Foundation

Gainesville, FL.

April 1993

Editors' Preface

The conference on Semigroups of Operators and Applications took place in Curaçao during August 1992. In the first week the two editors gave short courses to the Caribbean Mathematics Foundation, Gisele Goldstein's lectures being on nonlinear semigroups and Jerry Goldstein's being on linear semigroups. Those short courses sketched the theory and emphasized the applications, especially to PDE. The second week was devoted to an international conference, and lectures were given by some of the leading people in the field.

Curaçao is a paradise, and everyone, especially the undersigned, was astonished by the breathtaking beauty of the island. Jorge Martinez, the Director of the Caribbean Mathematical Foundation, is a masterful organizer. From his choice of hotel and conference location through the elegant coffee breaks to the conference banquet and other entertainment, he was the perfect host. In particular, he arranged for perfect weather throughout the meeting. We were all touched by Jorge's love for mathematics and for the Caribbean region. All the participants had a wonderful time.

The first two papers are written versions of the short courses given by the editors. The audience was lively and attentive and we are grateful for that. The rest of the papers are from the international conference portion of the program.

Some of the lectures given at the conference appear here as jointly authored papers. The corresponding lectures were presented by Alfonso Castro, Philippe Clément, Bill Fitzgibbon, Mary Parrott, and Michel Pierre.

Also, several interesting lectures were given which do not appear in these written proceedings. Those were:

J. Robert Dorroh: "Existence and regularity of solutions of singular quasi-linear diffusion equations",

Klaus Engel: "On dissipative wave equations in reflexive Banach spaces",

G. R. Goldstein and J. A. Goldstein (eds.),
Semigroups of Linear and Nonlinear Operations and Applications, 7–8.

Hernan Henriquez: "Periodic solutions of quasi-linear partial functional differential equations with unbounded delay",

Pedro Isaza J. and Jorge E. Mejia L. : "Local solutions of the Kadomstev-Petviashvili equation with periodic conditions",

Enzo Mitidieri: "Blow-up estimates for semilinear parabolic systems",

Frank Neubrander: "Degenerate abstract Cauchy problems",

Andreas Stahel: "Equations of a vibrating plate",

Silvia Teleman: "One subspace",

Ricardo Weder: "Obstacle scattering without local compactness".

The Caribbean and Latin Americans present at the conference were Halvard White of Jamaica, H. Antonius, C. Gorisson and Henrietta Ilahi of Surinam, P. Isaza and J. Mejia of Columbia, H. Henriquez and C. Lizama of Chile, I. Cioranescu and S. Teleman of Puerto Rico, and R. Weder of Mexico.

We are grateful to two members of the LSU staff who cheerfully and efficiently did much typing for us. Susan Oncal handled the correspondence and Jacquie Paxton did the technical typing.

Finally we are grateful to NSF for their support of our research.

Gisèle Ruiz Goldstein

Jerome A. Goldstein

Baton Rouge, LA.

April 1993

A Survey of Semigroups of Linear Operators and Applications

Jerome A. Goldstein

One parameter semigroups of linear operators have a simple elegant theory together with a wide variety of applications, covering a broad area. These lecture notes give an introduction to this theory and indicate some of the applications.

The emphasis here is on the extraordinary variety of applications. The presentation will be breezy. Proofs will usually be replaced by sketches or heuristic arguments.

Our starting point is the *abstract Cauchy problem* (or abstract initial value problem)

$$\frac{du}{dt} = Au \qquad (t \geq 0), \qquad u(0) = f. \qquad (ACP)$$

Here u maps $\mathbb{R}^+ = [0, \infty)$ into a space X, and the term "abstract" indicates that X is taken to be a Banach space, usually infinite dimensional. The meaning of $du/dt = Au$ is that

$$\frac{u(t+h) - u(t)}{h} - A(u(t)) \to 0$$

as $h \to 0$. Thus we want to take linear combinations and limits of members of X; taking X to be a Banach space is a natural assumption. The operator $A : \mathcal{D}(A) \subset X \to X$ is linear. The initial condition $u(0) = f$ makes sense if one views u as a continuous curve in X.

G. R. Goldstein and J. A. Goldstein (eds.),
Semigroups of Linear and Nonlinear Operations and Applications, 9–57.

The problem (ACP) is *well posed* if a solution *exists*, is *unique*, and *depends continuously* on the ingredients of the problem (namely f and A). The well-posed problems are the ones which interest us in this course.

Uniqueness allows us to express the solution at time $t+s$ in two ways. One is as $u(t+s) = T(t+s)f$, where $T(\tau)$ maps the solution at some time to the solution τ units of time later. ($T(\tau)$ only depends on the length τ of the time interval and not on its end points since the equation is autonomous, i.e., since A does not depend on the time t.) One can also write $u(t+s) = T(t)g$, where $g = T(s)f$ is the solution at time s. Thus

$$T(t+s)f = T(t)(T(s)f).$$

Since each $T(\tau)$ should be continuous and since we may assume that $\mathcal{D}(A)$ is dense in X (otherwise replace X by $\overline{\mathcal{D}(A)}$), we are led to the following definition. $T = \{T(t) : t \in \mathbb{R}^+\}$ is a *strongly continuous one parameter semigroup of operators* on X (or simply a (C_o) *semigroup* on X) if

$$T(t+s) = T(t)T(s) \text{ for all } t, s \geq 0,$$

$$T(0) = I,$$

$$T(\cdot)f \in C(\mathbb{R}^+, X) \text{ for all } f \in X,$$

i.e., $t \to T(t)f$ is continuous from $\mathbb{R}^+$ to X for all $f \in X$. Starting with a (C_o) semigroup T, one can associate an operator A and a corresponding (ACP) with T as follows. The (*infinitesimal*) *generator* of T is $A = T'(0)$. More precisely, f is in the domain of A, $\mathcal{D}(A)$, iff $\lim\limits_{h \to 0} \frac{T(h)f - f}{h}$ exists; then Af is the limit.

Formally $T(t) = e^{tA}$, and

$$\int_0^\infty e^{-\lambda t} T(t) dt = \int_0^\infty e^{-t(\lambda - A)} dt = (\lambda I - A)^{-1}. \qquad (1)$$

Suppose T is a (C_o) *contraction* semigroup, i.e. a (C_o) semigroup satisfying $||T(t)|| \leq 1$ for all $t \geq 0$. It follows from the above formal calculation that

$$\text{for all } \lambda > 0, \ ||(\lambda I - A)^{-1}|| \leq \frac{1}{\lambda} \qquad (2)$$

and

$$\mathbb{R}(\lambda I - A) = X \text{ for all } \lambda > 0. \qquad (3)$$

Here $\mathbb{R}(B)$ is the range of an operator B. Setting $\alpha = \frac{1}{\lambda,}$ (2) is equivalent to

$$\text{for all } \alpha > 0, \quad ||(I - \alpha A)^{-1}|| \leq 1. \qquad (2')$$

An operator satisfying (2) [or (2′)] is called *dissipative;* an operator satisfying (2) and (3) is called *m-dissipative.*

HILLE-YOSIDA THEOREM.

An operator A is the generator of a (C_o) contraction semigroup $T = \{T(t) : t \in \mathbb{R}^+\}$ iff $\overline{\mathcal{D}(A)} = X$ and A is m-dissipative. In this case (ACP) is well-posed; if $f \in \mathcal{D}(A)$, the unique solution $u \in C(\mathbb{R}^+, \mathcal{D}(A)) \cap C^1(\mathbb{R}^+, X)$ is given by $u(t) = T(t)f$. Finally,

(representation)

$$T(t)f = \lim_{n\to\infty} (I - \frac{t}{n} A)^{-n} f \ \textit{for all } t, f;$$

(regularity)

$$T(t)(\mathcal{D}(A^n)) \subset \mathcal{D}(A^n) \ \textit{for all } t \in \mathbb{R}^+ \textit{ and } n \in \mathbb{N} = \{1, 2, 3, \ldots\},$$

and if $f \in \mathcal{D}(A^n)$, $T(\cdot)f \in C(\mathbb{R}^+, \mathcal{D}(A^n)) \cap C^n(\mathbb{R}^+, X)$.

The proof is not difficult. The necessity follows easily from the variant of (1) with a vector in it to make the integrals exist:

$$\int_0^\infty e^{-\lambda t} T(t) f dt = (\lambda I - A)^{-1} f$$

for $f \in X$ and $\lambda > 0$. Thus $(\lambda I - A)^{-1}$ is the Laplace transform of the semigroup T. One way to prove the sufficiency is to invert the Laplace transform. Many other proofs are known.

The main examples come from partial differential equations (PDE):

heat equation $\partial u/\partial t = \Delta u$

wave equation $\partial^2 u/\partial t^2 = \Delta u$

Schrödinger equation $i\partial u/\partial t = \Delta u.$

Consider the heat equation for $t \geq 0$ with $x \in \Omega$, a smooth domain in $\mathbb{R}^n$. Associate with the Laplacian Δ the homogeneous Dirichlet boundary condition $u = 0$ for $x \in \partial\Omega$ ($=$ boundary of Ω) and all $t \geq 0$. Let X be a space of functions on Ω (e.g. $L^p(\Omega), 1 \leq p < \infty$). Let A be the (distributional) Laplacian acting on $\mathcal{D}(A) = \{f \in X \cap C(\bar{\Omega}) : \Delta f \in X, \quad f|_{\partial\Omega} = 0\}$. By identifying $x \to u(t,x)$ with $u(t) \in X$, the problems

$$\partial u/\partial t = \Delta u \text{ in } \mathbb{R}^+ \times \Omega,$$
$$u(x,0) = f(x) \text{ in } \Omega,$$
$$u(x,t) = 0 \text{ on } \mathbb{R}^+ \times \partial\Omega$$

and

$$du(t)/dt = Au(t) \qquad (t \in \mathbb{R}^+),$$
$$u(0) = f$$

are formally equivalent. In this case, either A or the closure $\bar{A}$ of

A is densely defined and m-dissipative. A similar result holds for the Schrödinger equation in L^2. To make the wave equation look like (ACP), simply reduce it to a first order system: Let

$$v = \begin{pmatrix} v_1 \\ v_2 \end{pmatrix} = \begin{pmatrix} u \\ \partial u/\partial t \end{pmatrix}, \qquad A = \begin{pmatrix} 0 & I \\ \Delta & 0 \end{pmatrix}.$$

Then $u_{tt} = \Delta u$ is formally equivalent to $dv/dt = Av$ on a space of pairs of functions on Ω. Notice that boundary conditions can be present in (ACP); they are built into the description of $\mathcal{D}(A)$.

SIMILARITY.

Let X, Y be Banach spaces. Let $U : X \to Y$ be linear, bijective and bicontinuous. Let $\widetilde{A} = UAU^{-1}$, $\widetilde{T}(t) = UT(t)U^{-1}$. Then A generates the (C_o) semigroup T on X iff $\widetilde{A}$ generates the (C_o) semigroup $\widetilde{T}$ on Y. An example of such a U is the Fourier transform F, which is unitary from $L^2(\mathbb{R}^n_x)$ to $L^2(\mathbb{R}^n_\xi)$. Thus, for $\widehat{f} = Ff$, $\widehat{f}(\xi) = (2\pi)^{-n/2} \int_{\mathbb{R}^n} e^{-ix\cdot\xi} f(x)dx$ (for $f \in L^2(\mathbb{R}^n_x) \cap L^1(\mathbb{R}^n_x)$). For $\partial^\alpha = \partial_1^{\alpha_1} \partial_2^{\alpha_2} \dots \partial_n^{\alpha_n}, \partial_j = \partial/\partial x_j$, $(\partial^\alpha f)^\wedge(\xi) = (i\xi)^\alpha \widehat{f}(\xi)$; thus F converts differential operators into multiplication operators. $|\alpha| = \sum_{j=1}^n \alpha_j$ is the order of ∂^α.

Let $X = L^2(\Omega, \Sigma, \mu)$ and let $m : \Omega \to \mathbb{C}$ be Σ-measurable. The operator of multiplication by m is M_m defined by $(M_m f)(\omega) = m(\omega) f(\omega)$, $\omega \in \Omega$, and $f \in \mathcal{D}(M_m)$ iff $f, mf \in X$. Let $m(\xi) = -c|\xi|^2$, $X = L^2(\mathbb{R}^n)$. For $X = L^2(\mathbb{R})$ let $\nu(\xi) = i\xi$. Then

$$c\Delta = F^{-1} M_m F, \qquad d/dx = F^{-1} M_\nu F,$$

$$e^{tc\Delta} = F M_{e^{tcm}} F, \qquad e^{td/dx} = F^{-1} M_{e^{t\nu}} F,$$

$$(e^{td/dx} f)(x) = f(x+t),$$

$$(e^{tc\Delta}f)(x) = (4\pi ct)^{-n/2}\int_{\mathbb{R}^n} e^{-|x-y|^2/4ct}f(y)dy. \tag{4}$$

When $c = 1$, this gives the formula for the solution of (ACP) for the heat equation on $\mathbb{R}^n$ with $t > 0$ (or $Re\, t > 0$). When $c = -i$, this gives the solution of (ACP) for the Schrödinger equation for $t \in \mathbb{R}$. In the latter case T is a (C_o) *unitary group* , i.e. each $T(t) = e^{-it\Delta}$ is a unitary operator on $L^2(\mathbb{R}^n)$. In the former case $\{T(t) : Re\, t \geq 0\}$ is analytic in the right half plane $\{Re\, t > 0\}$.

ANALYTIC SEMIGROUPS.

Let $\sum$ be the pictured shaded sector in the complex plane. Let A satisfy $\sum \subset \rho(A)$, the resolvent set of A, and $||(\lambda I - A)^{-1}|| \leq M/|\lambda|$ for some $M > 0$ and all $\lambda \in \sum$.

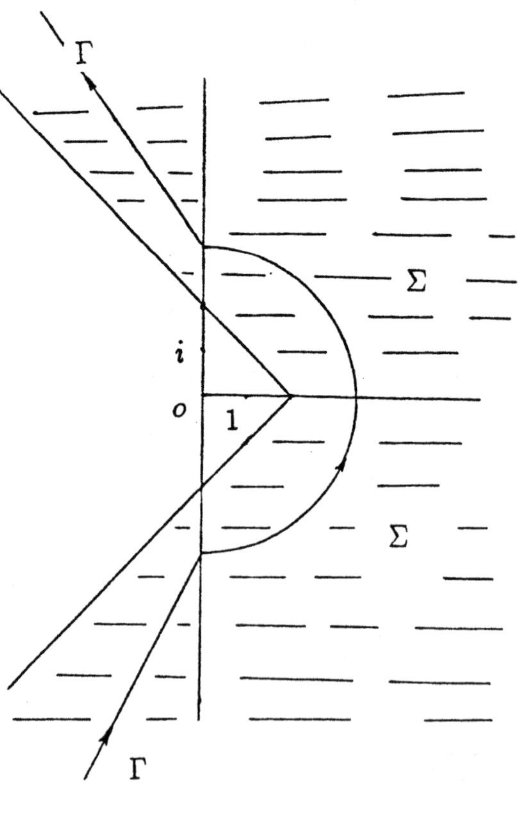

Then

$$T(t)f = \frac{1}{2\pi i}\int_\Gamma e^{\lambda t}(\lambda I - A)^{-1} f dt$$

defines a (C_o) semigroup (if $\overline{\mathcal{D}(A)} = X$) analytic in some sector including the positive real numbers. Moreover

$$T(t)(X) \subset \mathcal{D}(A^m) \text{ for all } m \in \mathbb{N}.$$

In the case of the heat equation on $\mathbb{R}^n$ (see (4) with $c = 1$), for all $f \in L^2(\mathbb{R}^n)$, $t > 0$,

$$u(t, \cdot) = T(t)f \in \bigcap_{m=1}^{\infty} \mathcal{D}(A^m) = \bigcap_{m=1}^{\infty} H^{2m}(\mathbb{R}^n) \subset C^\infty(\mathbb{R}^n),$$

and

$$u \in C^\infty\big((0,\infty) \times \mathbb{R}^n\big).$$

HILBERT SPACE RESULTS.

When $X = \mathcal{H}$ is a Hilbert space, A is dissipative iff $Re\langle Af, f\rangle \leq 0$ for all $f \in \mathcal{D}(A)$. [A similar result using semi-inner products or duality maps holds in a general Banach space.]

Let $B : \mathcal{D}(B) \subset \mathcal{H} \to \mathcal{H}$ be densely defined. Then $g \in \mathcal{D}(B^*)$ and $B^*g = h$ means: there is a constant C such that

$$|\langle Bf, g\rangle| \leq C||f||$$

holds for all $f \in \mathcal{D}(B)$. Then the bounded linear functional $f \to \langle Bf, g\rangle$ is given by $\langle f, h\rangle (= \langle f, B^*g\rangle)$. The denseness of $\mathcal{D}(B)$ implies that h is uniquely determined.

For $\mathcal{D}(A)$ dense in $\mathcal{H}$, A is *symmetric* if $\langle Afg\rangle = \langle f, Ag\rangle$ for all $f, g \in \mathcal{D}(A)$ iff $A \subset A^*$. A is *selfadjoint* if $A = A^*$, i.e. A is symmetric and $\mathcal{D}(A^*) = \mathcal{D}(A)$. The next two theorems can be proved as a consequence of the Hille-Yosida theorem. We take the Hilbert space to be complex.

STONE'S THEOREM.

A generates a (C_o) unitary group on $\mathcal{H}$ iff iA is selfadjoint.

SPECTRAL THEOREM.

(Hilbert, Stone, von Neumann). A on $\mathcal{H}$ is selfadjoint iff there is a measure space (Ω, Σ, μ), a unitary map $U : \mathcal{H} \to L^2(\Omega, \Sigma, \mu)$, and a Σ-measurable real function $m : \Omega \to \mathbb{R}$ such that

$$A = U^{-1} M_m U.$$

In this case, the spectrum $\sigma(A)$ of A is the closure of the (essential) range of m. For all measurable functions F from $\sigma(A)$ to $\mathbb{C}$, $F(A) = U^{-1} M_n U$ is well defined, where $n = F(m)$. In particular $e^{itA} = U^{-1} M_{e^{itm}} U$.

The wave equation on $\mathbb{R}^n$,

$$u_{tt} = \Delta u \quad (x \in \mathbb{R}^n, \quad t \in \mathbb{R}),$$

can be written as

$$\begin{pmatrix} u \\ u_t \end{pmatrix}_t = \begin{pmatrix} 0 & I \\ \Delta & 0 \end{pmatrix} \begin{pmatrix} u \\ u_t \end{pmatrix}.$$

Applying the Fourier transform F (in the space variables) gives

$$\begin{pmatrix} \widehat{u} \\ \widehat{u}_t \end{pmatrix}_t = \begin{pmatrix} 0 & 1 \\ -|\xi|^2 & 0 \end{pmatrix} \begin{pmatrix} \widehat{u} \\ \widehat{u}_t \end{pmatrix}.$$

A conserved quality is the *energy* E given by

$$\begin{aligned} E &= ||\widehat{u}_t||_{L^2}^2 + ||\ |\xi|\widehat{u}\ ||_{L^2}^2 \\ &= \langle \widehat{u}_t, \widehat{u}_t \rangle + \langle |\xi|^2 \widehat{u}, \widehat{u} \rangle = ||u_t||_{L^2}^2 + ||(-\Delta)^{1/2} u||_{L^2}^2 \\ &= ||u_t(t)||_{L^2}^2 + ||\nabla_x u(t)||_{L^2}^2. \end{aligned}$$

The fact that $dE/dt \equiv 0$ leads to the conclusion that the wave equation is governed by a (C_o) unitary group on

$$\mathcal{H} = \{\begin{pmatrix} w_1 \\ w_2 \end{pmatrix} : ||w_2||_{L^2}^2 + ||(-\Delta)^{1/2} w_1||_{L^2}^2 < \infty\};$$

and $E^{1/2}$ is called the *energy norm* of the solution u.

Roughly speaking, *parabolic equations* (such as the heat equation) are governed by analytic semigroups. *Hyperbolic equations* (such as the wave equation) are governed by groups, i.e., they are well-posed in forward time and in backward time. From this point of view, the Schrödinger equation is hyperbolic. Some equations (such as many functional differential equations) are neither parabolic nor hyperbolic; they are governed by (C_o) semigroups which are neither analytic nor groups.

PERTURBATION THEOREMS.

I. *Let A generate a (C_o) contraction semigroup on X, i.e., A is densely defined and m-dissipative. Let B be dissipative with $\mathcal{D}(B) \supset \mathcal{D}(A)$ and suppose these exist constants $a < 1$, $b \geq 0$ such that*

$$||Bf|| \leq a||Af|| + b||f|| \tag{5}$$

holds for each $f \in \mathcal{D}(A)$. Then $A + B$ (defined on $\mathcal{D}(A)$) is m-dissipative.

This is a variant of a perturbation theorem about selfadjoint operators, due to Kato and Rellich.

II. *Let A be selfadjoint on $\mathcal{H}$. Let B be symmetric with $\mathcal{D}(B) \supset \mathcal{D}(A)$ and suppose that (5) holds, with $a < 1$. Then $A + B$ (on $\mathcal{D}(A)$) is selfadjoint; and $A + B$ is semibounded if A is.*

III. *Let A generate an analytic semigroup on X and let $\mathcal{D}(B) \supset \mathcal{D}(A)$. Suppose that there is a sufficiently small $a > 0$ and some $b > 0$ such that (5) holds. Then $A + B$ (on $\mathcal{D}(A)$) generates an analytic semigroup on X.*

The "sufficiently small" part of the statement of Theorem III is vague. In the applications, one can usually show that given any $\varepsilon > 0$, there is a $C_\varepsilon > 0$ such that one can take $a = \varepsilon, b = C_\varepsilon$ in Theorems I, II, or III.

Here is an example. Let $A = \Delta$ or more generally let $A = \sum_{|\alpha|=|\beta|=m} a_{\alpha\beta}\partial^{\alpha+\beta}$. Here each $a_{\alpha\beta}$ is real and $a_{\alpha\beta} = a_{\beta\alpha}$ is assumed. On $\mathcal{H} = L^2(\mathbb{R}^n)$,

$$\langle Au, u\rangle = \sum_{|\alpha|=|\beta|=m} (-1)^m a_{\alpha\beta} \int_{\mathbb{R}^n} \partial^\alpha u \, \overline{\partial^\beta u} \, dx.$$

To insure that this is nonpositive we assume the *ellipticity hypothesis:* There is an $\varepsilon_o > 0$ such that

$$\sum_{|\alpha|=|\beta|=m} (-1)^m a_{\alpha\beta}\xi^\alpha\xi^\beta \leq -\varepsilon_0|\xi|^{2m}$$

for each $\xi \in \mathbb{R}^n$. Here $\xi^\alpha = \xi_1^{\alpha_1}\xi_2^{\alpha_2}\ldots\xi_n^{\alpha_n}$, as usual. Thus Δ is elliptic with $m = 1$, $a_{jk} = \delta_{jk}$, and $\varepsilon_0 = 1$. By the spectral theorem and the associated operational calculus, A is a nonpositive selfadjoint operator and A generates a semigroup analytic in the right half plane. The last sentence holds for $A + B$ as well provided

$$B = \sum_{0\leq|\alpha|\leq 2m-1} b_\alpha(x)\partial^\alpha$$

where each $b_\alpha \in L^\infty(\mathbb{R}^n)$. The proof is as follows. Given $k \in \{0, 1, \ldots, 2m-1\}$, for each $\varepsilon > 0$ there is a $C = C(\varepsilon; 2m, n) > 0$

such that

$$\sum_{|\alpha|=k} ||\partial^\alpha f||_{L^2}^2 \leq \varepsilon ||Af||_{L^2}^2 + C||f||_{L^2}^2.$$

This is easy to see using Fourier transforms together with

$$\max_{0 \leq k \leq 2m-1} |\xi|^k \leq \varepsilon |\xi|^{2m} + C(\varepsilon; 2m, n),$$

which holds for all $\xi \in \mathbb{R}^n$. Since each b_α is bounded, Perturbation theorem III applies to finish the proof.

The same results holds if $\mathbb{R}^n$ is replaced by $\Omega \subset \mathbb{R}^n$. In this case one must impose boundary conditions to make A selfadjoint and nonnegative. More generally A can have variable coefficients (i.e., the $a_{\alpha\beta}$ can depend on $x \in \Omega$) and one can work in the space $L^p(\Omega)$, $1 < p < \infty$. Solving $u - \lambda Au = h$ (or $u - \lambda(A+B)u = h$) and estimating in terms of λ and h is done using the theory of elliptic boundary value problems.

We summarize. It is easy to construct examples of parabolic PDEs governed by analytic semigroups on $X = L^2(\Omega)$, with $\Omega = \mathbb{R}^n$, A a constant coefficient elliptic operator and B an operator with lower order terms and bounded coefficients. Analogous extentions to $\Omega \subset \mathbb{R}^n$, $X = L^p(\Omega)$ and variable coefficients for A all hold but are technically much more complicated and rely on deep facts from linear elliptic PDE theory.

Similar results hold for wave equations of the form

$$u_{tt} + (A + B)u = 0$$

using Perturbation theorem II, with $A = \sum_{|\alpha|=|\beta|=m} a_{\alpha\beta}\partial^{\alpha+\beta}$ as before and with

$$Bu = \sum_{|\alpha|,|\beta| \le m-1} (-1)^{|\alpha|+|\beta|} \partial^\alpha \left[b_{\alpha\beta}(x) \partial^\beta u \right]$$

with b_{α_β} complex valued, $b_{a_\beta} = \bar{b}_{\beta\alpha}$, and $\partial^\gamma b_{\alpha_\beta} \in L^2(\mathbb{R}^n)$ for each γ with $|\gamma| \le \max\{|\alpha|, |\beta|\}$. This includes

$$u_{tt} = \Delta u + \sum_{j=1}^{n} b_j(x) \partial u / \partial x_j + c(x) u$$

for $x \in \mathbb{R}^n$, $t \in \mathbb{R}$ for certain choices of $b_1, \ldots, b_n, c$.

If $C = A + B$ is positive and selfadjoint, then the energy norm associated with the unitary group governing $u_{tt} + Cu = 0$ is $E^{1/2}$, where

$$E = ||u_t(t)||^2 + ||C^{1/2} u(t)||^2.$$

APPROXIMATION THEOREM.

Let A_n generate a (C_o) contraction semigroup T_n on X for $n = 0, 1, 2, \ldots$. Then $T_n(t)f \to T_0(t)f$ holds for all $t \ge 0$ and all f in X iff $(\lambda - A_n)^{-1} f \to (\lambda - A_0)^{-1} f$ holds for all $\lambda > 0$ and all $f \in X$. Sufficient (but not necessary) is that $D(A_0) \subset \mathcal{D}(A_n)$ and $A_n f \to A_0 f$ holds for all $f \in \mathcal{D}(A_0)$.

A simple special case is as follows: Let $A_n \in \mathcal{B}(X)$ (the bounded linear operators on X) and A_n be dissipative for all $n \ge 1$, $A_n f \to A_0 f$ for all $f \in \mathcal{D}$, where $\mathcal{D}$ is a core for A_0, i.e. A_0 is the closure of its restriction to $\mathcal{D}$. Then the above conclusion holds. We give a simple proof of this in the event that A_n and $T_0(t)$ commute for all $t \ge 0$ and all $n \ge 1$. For $f \in \mathcal{D}$,

$$\begin{aligned}
||T_n(t)f - T_0(t)f|| &= || - \int_0^t \frac{d}{ds} T_n(t-s)T_0(s)f ds|| \\
&= || \int_0^t T_n(t-s)(A_n - A_0)T_0(s)f ds|| \\
&\le || \int_0^t ||T_n(t-s)T_0(s)(A_n - A_0)f|| ds \\
&\text{by the commutativity assumption} \\
&\le \int_0^t ||(A_n - A_0)f|| ds \\
&= t||A_n f - A_0 f|| \to 0
\end{aligned}$$

as $n \to \infty$, uniformly for t in bounded sets in $\mathbb{R}^+$. This last conclusion holds in general in the context of the approximation theorem.

EXAMPLE.

Let $X = BUC(\mathbb{R})$, the bounded uniformly continuous functions on $\mathbb{R}$, which is a Banach space under the supremum norm.

$$A_0 = d/dx \text{ with } \mathcal{D}(A_0) = \{f \in X : f \in C^1(\mathbb{R}), f' \in X\}$$

generates the (C_o) contraction semigroup T_0 given by

$$\big(T_0(t)f\big)(x) = f\big(x+t\big).$$

(Cf. (4).) Let $A_n = \Big[\big(T(\tfrac{1}{n}) - I\big)/(1/n)\Big]$, for $n \in \mathbb{N}$. Then, uniformly for $x \in \mathbb{R}$ and $t \in [0,1]$,

$$f(x+t) = \lim_{n\to\infty} \lim_{M\to\infty} \sum_{m=0}^{M} \frac{t^m \left(A_n^m f\right)(x)}{m!}.$$

Letting $x = 0$ yields

$$f(t) = \lim_{n\to\infty} \sum_{m=0}^{M_m} t^m C_{mn}$$

with $C_{mn} = (A_n^m f)(0)/m!$ and M_m suitably chosen; this holds uniformly. Thus any $f \in C[0,1]$ (which can be trivally extended to be in X) is a uniform limit of polynomials; this is nothing but the Weierstrass approximation theorem. A simple modification of this argument yields a deeper and more interesting result due to Hille. Consider the difference operator Δ_h defined by

$$(\Delta_h f)(x) = \frac{f(x+h) - f(x)}{h}.$$

Then the approximation theorem (i.e. the version with $h \to 0$ replacing $n \to \infty$) yields

$$f(x+t) = \lim_{h \to 0} \lim_{M \to \infty} \sum_{m=1}^{M} \frac{t^m}{m!} (\Delta_h^m f)(x), \tag{6}$$

uniformly for $x \in \mathbb{R}$ and t is bounded intervals, for all $f \in X$. Now look at (6) with the limits reversed:

$$f(x+t) = \lim_{M \to \infty} \lim_{h \to 0} \sum_{m=0}^{M} \frac{t^m}{m!} (\Delta_h^m f)(x).$$

This is of course Taylor's theorem and it has many hypotheses, the main one being that f is *analytic* on $\mathbb{R}$. But the form (6) works without *any* differentiability hypothesis! If one wants $\sum_{m=0}^{M} \frac{t^m}{m!} (\Delta_h^m f)(x)$ to be a good approximation to $f(x+t)$ for general continuous f, one should fix $h \neq 0$ and take many terms; that is M should get large before h gets small.

CHERNOFF FORMULA.

Let $V(\cdot) : \mathbb{R}^+ \to \mathcal{B}(X)$ satisfy $V(0) = I$, $||V(t)|| \leq 1$ for each $t \geq 0$, $V(\cdot)f \in C(\mathbb{R}^+, X)$ for each $f \in X$, and $V'(0) = A$ on $\mathcal{D}$ where $\overline{A|_{\mathcal{D}}} = A$ generates a (C_o) semigroup T. (The last *condition can be restated as* $\frac{V(h)f-f}{h} \to Af$ *for all* f *in a core* $\mathcal{D}$ *of* A*.) Then*

$$\lim_{n\to\infty} V\left(\frac{t}{n}\right)^n f = T(t)f$$

for all $t \geq 0$ *and* $f \in X$*.*

The convergence is uniform for t in bounded sets, but that will not concern us here.

Lemma. *Let* $L \in \mathcal{B}(X)$ *with* $||L|| \leq 1$*. Then for all* $\in X$*,*

$$||e^{-n(L-I)}f - L^n f|| \leq \sqrt{n}||Lf - f||.$$

We omit the elementary proof.

Now let the hypotheses of the Chernoff formula hold and take $L = V(t/n),\ A_n = \frac{V(\frac{t}{n})-I}{\frac{t}{n}}$. Then for $\alpha > 0$,

$$||e^{\alpha(L-I)}|| = e^{-\alpha}||e^{\alpha L}|| \leq e^{-\alpha}e^{\alpha||L||} \leq 1$$

since $||L|| \leq 1$. Thus A_n is m-dissipative. For $f \in \mathcal{D}$,

$$V\left(\frac{t}{n}\right)^n f - T(t)f = \{L^n f - e^{n(L-I)}f\}$$
$$+ \{e^{n(L-I)}f - T(t)f\} \equiv J_1 + J_2.$$
$$||J_1|| \leq \sqrt{n}||Lf - f|| = \frac{t}{n}\sqrt{n}||\frac{V(\frac{t}{n})f - f}{t/n}|| \to 0$$

as $n \to \infty$ by the lemma, since the above expression approximately equals

$$\frac{t}{\sqrt{n}}||Af||.$$

Next,

$$||J_2|| = ||e^{tA_n}f - T(t)f|| \to 0$$

as $n \to \infty$ by the approximation theorem, since $A_n f \to Af$ on $\mathcal{D}$. The Chernoff formula follows. □

CENTRAL LIMIT THEOREM.

A *probability space* is a measure space (Ω, Σ, P) satisfying $P(\Omega) = 1$. A *random variable* is a Σ-measurable function $\xi : \Omega \to \mathbb{R}$. The *distribution function* F of ξ is given by

$$F_\xi(t) = P[\xi \leq t] = P\{\xi^{-1}(-\infty, t]\}, \qquad t \in \mathbb{R}.$$

Let $\mathcal{DF}$ be the set of all distribution functions, i.e., the set of all monotone nondecreasing $F : \mathbb{R} \to \mathbb{R}$ which are continuous from the right and satisfy $\lim_{t\to-\infty} F(t) = 0$, $\lim_{t\to\infty} F(t) = 1$. Let $X = BUC(\mathbb{R})$, the bounded uniformly continuous functions on $\mathbb{R}$. Define the "tilde transform" to be the map

$$F \to \widetilde{F}$$

from $\mathcal{DF}$ to $\mathcal{B}(X)$ given by

$$(\widetilde{F}f)(t) = \int_{-\infty}^{\infty} f(t-s)dF(s), \quad f \in X, t \in \mathbb{R}.$$

An abbreviated notation is $\widetilde{F}f = f * dF$.

Lemma.

i) *$\widetilde{F}_n \to \widetilde{F}_0$ in the strong operator topology iff $F_n \to F_0$ in distribution. That is, for $F_n \in \mathcal{DF}$, $||\widetilde{F}_n f - \widetilde{F}_0 f|| \to 0$ for all $f \in X$ iff $F_n(x) \to F_0(x)$ for all $x \in \mathbb{R}$ at which F_0 is continuous.*

ii) $(F * G)\widetilde{\ } = \widetilde{F}\widetilde{G}$.

The elementary proof is omitted. Note that i) says that $F \to \widetilde{F}$ is bicontinuous; ii) says that it converts convolution to ordinary multiplication (or composition) of linear operators.

This is significant since for independent random variables the distribution function of the sum is the convolution of the distribution functions.

A sequence $\xi = \{\xi_1, \xi_2, \ldots, \xi_n, \ldots,\}$ of random variables is called *independent* if for every finite subset $\{n_1, \ldots, n_k\}$ of $\mathbb{N}$ and for all $A_k \in \Sigma$, $\quad P\{\xi_{n_i} \in A_i \text{ for } 1 \leq i \leq k\} = \prod_{i=1}^{k} P\{\xi_{n_i} \in A_i\}$. Equivalent is that $F_{\sum_{i=1}^{k} \xi_{n_i}} = F_{\xi_{n_1}} * F_{\xi_{n_2}} * \ldots * F_{\xi_{n_k}}$. The sequence ξ is called *identically distributed* if $F = F_{\xi n}$ is independent of n.

CENTRAL LIMIT THEOREM.

Let $\xi_i, \xi_2, \ldots$ be independent identically distributed random variables with mean 0 and variance 1. Then as $N \to \infty$, the normalized sum $\frac{1}{\sqrt{N}} \sum_{i=1}^{N} \xi_i$ converges in distribution to a standard normal random variable.

Some of these terms must be explained . For Borel functions g on $\mathbb{R}$,

$$E(g(\xi_i)) = \int_{\Omega} g(\xi_i) dP = \int_{-\infty}^{\infty} g(x) dF_{\xi_i}(x),$$

with the understanding that whenever one integral exists, they both do and equality holds. The most important numbers of this form are the *mean*

$$\mu = E(\xi_i) = \int_{-\infty}^{\infty} x dF(x)$$

and the *variance*

$$\sigma^2 = E\Big((\xi_i - \mu)^2\Big) = \int_{-\infty}^{\infty} (x - \mu)^2 dF(x).$$

Here we drop the subscript i from F since the ξ_i are identically distributed; the two functions used here are $g(x) \equiv x$ and $g(x) \equiv$

$(x-\mu)^2$. If ξ has a finite variance, i.e. if $\xi \in L^2(\Omega, \Sigma, P)$, then replacing ξ_i by $\eta_i = \frac{\xi_i - \mu}{\sigma}$ gives that η_i has mean 0 and variance 1. This holds without loss of generality for nonconstant random variables (i.e. $\sigma > 0$). The distribution function for the standard normal distribution is

$$F_{N(0,1)}(x) = \frac{1}{\sqrt{2\pi}} \int_{-\infty}^{x} e^{-t^2/2} dt.$$

Thus the conclusion of the Central Limit theorem is that for all $x \in \mathbb{R}$,

$$\lim_{N\to\infty} F_{\frac{1}{\sqrt{N}} \sum_{i=1}^{N} \xi_i}(x) = \frac{1}{\sqrt{2\pi}} \int_{-\infty}^{x} e^{-t^2/2} dt.$$

Note that $\eta = \frac{1}{\sqrt{N}} \sum_{i=1}^{N} \xi_i$ is the normalized sum of $\xi_1, \ldots, \xi_N$ in that η has mean 0 and variance 1.

Here is a quick outline of the proof.

Let $X = BUC(\mathbb{R})$,

$$D = \{f \in X \cap C^2(\mathbb{R}) : f', f'' \in X\},$$

$G = F_{\xi_i}$ the common distribution function and for $r > 0$ let $G_r(x) = G(\sqrt{r}x)$ for $x \in \mathbb{R}$. If $V(t) = \widetilde{G}_{1/t}$ with $V(0) = I$, then the hypotheses of Chernoff's theorem hold. A Taylor's theorem argument gives $V'(0)f = \frac{1}{2}f''$ for $f \in D$. Chernoff's formula implies

$$V(\frac{t}{n})^n f \to T(t)f.$$

But $V\left(\frac{t}{n}\right) = \widetilde{F}_\eta$ with $\eta = \sqrt{\frac{t}{n}} \sum_{i=1}^{n} \xi_i$; and since $A = \frac{1}{2}d^2/dx^2$, $T(t) = \widetilde{F}_{N(0,t)}$, i.e., the semigroup governing the heat equation (with the factor 1/2) is given by convolution with the distribution function of the normal distribution with mean 0 and variance t. The result follows with the aid of the lemma.

LIE-TROTTER-DALETSKII PRODUCT FORMULA.

Let A, A_2 and $A_3 = \overline{A_1 + A_2}$ be densely defined m-dissipative operators on X. Let T_i be the semigroup generated by A_i. Then

$$T_3(t)f = \lim_{n\to\infty} \left[T_1(\frac{t}{n})T_2(\frac{t}{n})\right]^n f$$

holds for all $t \geq 0$ and all $f \in X$.

This follows from Chernoff's formula. Take $V(t) = T_1(t)T_2(t)$ and calculate $V'(0) = A_1 + A_2$ on $\mathcal{D}(A_1) \cap \mathcal{D}(A_2)$.

FEYNMAN PATH FORMULA.

The *wave function* of a (nonrelativistic) quantum mechanical system in the solution $u = u(x,t)$ of the Schrödinger equation

$$i\frac{\partial u}{\partial t} = -\frac{1}{2m}\Delta u + V(x)u \quad (x \in \mathbb{R}^\ell, \qquad t \in \mathbb{R}).$$

$$u(x,0) = f(x) \quad (x \in \mathbb{R}^\ell).$$

Imagine N identical particles (say electrons), each of mass m. Then $\ell = 3N$ and the real function V is the potential describing the forces on the particles. It is assumed that the closure H of $\frac{-\Delta}{2m} + V$ (on $\mathcal{D}(\Delta) \cap \mathcal{D}(M_V)$) is selfadjoint; this is a very mild restriction on V. By Stone's theorem, $-iH$ generates a (C_o) unitary group U, and $u(t) = U(t)f$ for t real. The datum f is normalized to be a unit vector in $L^2(\mathbb{R}^\ell)$; then $||u(t)||_2 = 1$ for all real t. $|u(x,t)|^2$ [resp. $|\widehat{u}(\xi,t)|^2$] is the *position* [resp. *momentum*] probability density of the system. That is, $\int_\Gamma |u(x,t)|^2 dx$ is the probability that at time t the position vector of the system is in $\Gamma \subset \mathbb{R}^\ell$. Both $|u(\cdot,t)|^2$ and $|\widehat{u}(\cdot,t)|^2$ are probability densities because of the unitarity of the Fourier transform on $L^2(\mathbb{R}^\ell)$.

Apply the Lie-Trotter-Daletskii product formula with $A = i\Delta/2m, B = M_{-iV}$ and $C = \overline{A+B} = -iH$. We have

$$(T_A(t)f)(x) = (2\pi\, i\, t/m)^{-l/2} \int_{\mathbb{R}^l} e^{i\, m\, |x-y|^2/2t} f(y)dy,$$
$$(T_B(t)f)(x) = \exp\{-i\, tV(x)\}f(x).$$

If $S_n(t) = (T_A(t/n)T_B(t/n))^n$, then by induction,

$$(S_n(t)f)(x_0) = \left(\frac{2\pi\, i\, t}{n\, m}\right)^{-\ell/2} \underbrace{\int_{\mathbb{R}^\ell} \cdots \int_{\mathbb{R}^\ell}}_{n} e^{i\, S} f(x_n)dx_1 \cdots d_{x_n} \tag{7}$$

where

$$S = \sum_{j=1}^{n} \left[\frac{m}{2}\frac{|x_j - x_{j-1}|^2}{(t/n)^2} - V(x_j)\right]\frac{t}{n}.$$

Define the space Ω_x of continuous paths starting at x:

$$\Omega_x = \{\omega \in C(\mathbb{R}^+, \mathbb{R}^\ell) : \omega(0) = x\}.$$

If $x_j = \omega(jt/n)$ and $x_0 = x$, then S is a Riemann sum approximating the *action integral*

$$S(\omega; t) = \int_0^t \left[\frac{m}{2}|\dot{\omega}(s)|^2 - V(\omega(s))\right] ds.$$

Formally

$$(S_n(t)f)(x) \to C \int_{\Omega_x} e^{iS(\omega;t)} f(\omega(t))\mathcal{D}\omega. \tag{8}$$

This is the celebrated *Feynman integral formula* for the wave function $u(x,t)$. There are at least three mysterious aspects of it. First, the "constant" C satisfies $|C| = \lim_{n\to\infty} \left|\frac{2\pi\, it}{nm}\right|^{-\ell/2} = \infty$. Next, the "measure" $\mathcal{D}w = \prod_{0\leq s\leq t} dx_s \left(= \lim_{n\to\infty} dx_1 \dots dx_n\right)$ cannot be defined as a countably additive set function. Kac considered the associated heat equation $u' + Hu = 0, u(0) = f$ and represented the solution in a similar way as a Wiener integral over Ω_x. For Wiener integrals, paths ω which are somewhere differentiable form a null set. In other words, if one could interpret (8) as a Wiener intergral, then the integrand involves $S(\omega; t)$, which for $t > 0$ is infinite almost everywhere. Thus in the mysterious Feynman formula (8) neither the constant nor the measure nor the integrand makes sense. But the formula

$$u(x,t) = C \int_{\Omega_x} e^{i\,S(\omega,t)} f(\omega(t))\mathcal{D}\omega \tag{9}$$

has great insuitive appeal. It expresses a basic object of quantum mechanics as an average over paths involving classical mechanics notions, namely, the action. Feynman used this idea to great profit in his Nobel Prize winning work on quantum electrodynamics and elsewhere.

It is now easy to give a rigorious interpretation of (9). Namely, replace (9) by the Riemann like approximant (7). Then by the Lie-Trotter- Daletskii product formula, (8) holds, in the sense that $S_n(t)f$ converges in $L^2(\mathbb{R}^\ell)$ to the solution $u(\cdot, t)$.

The beautiful argument is due (independently) to E. Nelson and Y. Daletskii.

INHOMOGENEOUS EQUATIONS.

Consider

$$u' = Au + h, \quad u(0) = f$$

where A generates a (C_o) semigroup T on X. The *variation of parameters formula* represents the solution (if it exists) as

$$u(t) = T(t)f + \int_0^t T(t-s)h(s)ds.$$

If one regards $T(t)$ as e^{tA}, this is exactly the result of one variable calculus for the ODE $u' = au + h(t)$. The moral is that you can make a living in this subject if you are really proficient in one variable calculus.

The semilinear problem

$$u'(t) = Au(t) + g(t, u(t)), \quad u(0) = f$$

can be attacked by successive approximations of the form

$$u_n'(t) = Au_n(t) + g(t, u_{n-1}(t)), \; u_n(0) = f.$$

The solution of this linear inhomogeneous problem is, by the above, given by

$$u_n(t) = T(t)f + \int_0^t T(t-s)g(s, u_{n-1}(s))ds.$$

Thus solving our problem reduces to solving the integral equation

$$u(t) = T(t)f + \int_0^t T(t-s)g(s,\, u\,(s))ds.$$

If $(Su)(t)$ is the right hand side of the above equation, then one wishes to find a fixed point of S, viewed as a mapping from a subset of $C\big([0,\tau], X\big)$ into itself. Such a fixed point is called a *mild solution* of $u' = Au + g(t,u)$, $u(0) = f$. It may not be differentiable (depending on g), but if it is it is a solution in the usual sense. Also, in showing that S has a fixed point, one may have to take τ be be small (and positive), thus one gets a local solution. An additional *a* priori estimate is needed to get global existence, which does not always hold.

THE NAVIER-STOKES EQUATIONS.

Let $u(t,x) \in \mathbb{R}^3$ be the fluid velocity and $p(t,x) \in \mathbb{R}$ the pressure of a fluid occupying a (smooth bounded) container $\Omega \subset \mathbb{R}^3$. Here $x \in \Omega$ and $t \geq 0$ is the time. The motion of the (incompressible) fluid is given by the *Navier-Stokes equations,*

$$(NS)_0 \quad \begin{cases} u_t - \nu\Delta u = -(u\cdot\nabla)u - \nabla p + g & in[0,T]\times\Omega, \\ \text{div } u = 0 & in[0,T]\times\Omega, \\ u = 0 & on[0,T]\times\partial\Omega, \\ u = u_0 & on\{0\}\times\Omega. \end{cases}$$

This can be rewritten as

$$\frac{\partial u_j}{\partial t} - \nu\Delta u_j = -\sum_{k=1}^{3} u_k \frac{\partial u_j}{\partial x_k} - \frac{\partial p}{\partial x_j} + g_j, \qquad j = 1,2,3,$$

$$\sum_{h=1}^{3} \partial u_k/\partial x_k = 0,$$

$$u(t,x) = 0 \quad (t \geq 0, \quad x \in \partial\Omega),$$

$$u(0,x) = u_0(x) \quad (x \in \Omega).$$

Define the spaces

$$\mathcal{H} \equiv L^2(\Omega;\mathbb{C}^3) = \left[L^2(\Omega)\right]^3,$$

$$\mathcal{H}_\sigma \equiv c\ell_{\mathcal{H}}\{uC^1(\bar\Omega;\mathbb{C}^3) : \text{ div } u = 0 \text{ in } \Omega,\ u = 0 \text{ on } \partial\Omega\},$$

$$\mathcal{H}_\nabla \equiv c\ell_{\mathcal{H}}\{\nabla\varphi : \varphi \in C^1(\bar\Omega)\}.$$

Here $c\ell_{\mathcal{H}}$ means the closure is $\mathcal{H}$ and σ stands for solenoidal (or divergence free) vectors.

Lemma. $\mathcal{H} = \mathcal{H}_\sigma \oplus \mathcal{H}_\nabla$.

This observation, which goes back to Helmholtz, is often called the Hodge decomposition. To prove the orthogonality, note that

$$\langle u, \nabla\varphi\rangle = \int_\Omega u \cdot \nabla\varphi dx = \int_\Omega \{ \operatorname{div} (\varphi u) - \varphi \operatorname{div} u\} dx$$
$$= \int_{\partial\Omega} \varphi u \cdot \mathbf{n} dS = 0$$

by the divergence theorem where $\mathbf{n}$ is the unit outer normal and the fact that $\operatorname{div} u = 0$. It remains to write any $u \in \left[C_c^\infty(\Omega)\right]^3$ as $u = v + \nabla\varphi$ where $\operatorname{div} v = 0$. Let $h = \operatorname{div} u$. From $\operatorname{div} u = \operatorname{div} v + \operatorname{div} \nabla\varphi$, it follows that $\Delta\varphi = h$ and $\partial\varphi/\partial\mathbf{n} = \nabla\varphi\cdot\mathbf{n} = u\cdot\mathbf{n} = 0$ on $\partial\Omega$. Solving this Neumann problem gives φ; $v = u - \nabla\varphi$ gives the lemma. □

Let $P : \mathcal{H} \to \mathcal{H}_\sigma$ be the orthogonal projection onto $\mathcal{H}_\sigma$. The *Stokes operator* is

$$A = \nu P \Delta I_{3\times3} = \nu P \begin{pmatrix} \Delta & 0 & 0 \\ 0 & \Delta & 0 \\ 0 & 0 & \Delta \end{pmatrix}$$

where Δ is the Dirichlet Laplacian : $\mathcal{D}(\Delta) = H^2(\Omega) \cap H_0^1(\Omega)$. The Stokes operator is morally like the Dirichlet Laplacian.

Lemma. *$A = A^* \leq -\epsilon I$ on $\mathcal{H}$ for some $\varepsilon > 0$. A has an orthonormal basis of eigenvectors.*

Let $v = Pu$ (which is u if u is $\mathcal{H}_\sigma$ valued). Then $(NS)_0$ becomes

$$(NS)_1 \qquad \begin{cases} v'(t) = Av(t) + N(v(t)) + h(t), \\ v(0) = v_0. \end{cases}$$

Here $v_0 = Pu_0 = u_0$, $h = Pg$, $N(v) = -P(v \cdot \nabla)v$. The point is that $P(\nabla p) = 0$. Thus $(NS)_1$ is an equation in $\mathcal{H}_\sigma$. After solving

$(NS)_1$ uniquely for v, set $u = v$ and plug into $(NS)_0$. This gives ∇p, which determines p uniquely up to an additive constant.

Following earlier work of Leray, Hopf, and others, Kato and Fujita established local wellposedness of $(NS)_1$; i.e. they found a solution for $t \in [0, \tau]$ for some $\tau > 0$. Here is a outline of the argument.

Let $\mathcal{K}$ be a Hilbert space and A_0 a selfadjoint operator on $\mathcal{K}$ satisfying $A_0 = A_0^* \leq -\varepsilon I$. (Think of $\mathcal{K} = \mathcal{H}_0$ and $A_0 = A$.) By the spectral theorem, A_0 is (in some representation) multiplication by a function with values in $(-\infty, -\varepsilon]$ on some L^2 space. So define $\mathcal{K}_\alpha = \mathcal{D}((-A)^\alpha)$ with norm $||f||_\alpha = ||(-A_0)^\alpha f||$. Let $0 \leq \alpha < 1$ and let F map an open set containing f in $\mathcal{K}_\alpha$ to $\mathcal{K}$ and satisfy

$$||Ff_1 - Ff_2|| \leq K||f_1 - f_2||_\alpha$$

for some $K = K_f$ and all f_1, f_2 in the open set. Then there is a unique mild solution of $v' = Av + F(v), \quad v(0) = f$ in $C([0, \tau], X)$ for some $\tau > 0$. Since $x^\alpha e^{-\alpha} \leq 1$ for $x > 0$, the spectral theorem implies $||(-A_0)^\alpha e^{tA_0}|| \leq t^{-\alpha}$. Hence

$$\begin{aligned} &||\int_0^t e^{(t-s)A_0}(F(v_1(s)) - F(v_2(s)))ds|| \\ \leq &\int_0^t ||e^{(t-s)A_0}||_{\mathcal{B}(\mathcal{K},\mathcal{K}_\alpha)}||F(v_1(s)) - F(v_2(s))||_{\mathcal{K}_\alpha} ds \\ &\leq K \int_0^t (t-s)^{-\alpha}||v_1(s) - v_2(s)||_\alpha ds, \end{aligned}$$

and the integral exists for $t > 0$ since $(t-s)^{-\alpha}$ is integrable on $[0, t]$ because $\alpha < 1$. The Banach fixed point theorem (alias the Picard iteration method) gives the desired local solution.

To apply this result to $(NS)_1$, one must use Sobolev inequalities; the end result is that any $\alpha \in (3/4, 1)$ will suffice.

The problem of global wellposedness of $(NS)_1$ is still an open problem. Its solution will explain turbulence.

A SEMILINEAR PROBLEM WITH A NONLINEAR BOUNDARY CONDITION.

Here we give another example of solving a nonlinear problem by solving linear inhomogeneous problems.

Let $X = L^p(\Omega)$ where Ω is a smooth bounded domain in $\mathbb{R}^n$. Let $\mathcal{A}$ be the Laplacian Δ with domain $\mathcal{D}(\mathcal{A}) = W^{2,p}(\Omega)$. We have $W^{2,p}(\Omega) \hookrightarrow C(\bar{\Omega})$ provided $\max\{\frac{n}{2}, 1\} < p < \infty$, which we assume. Let Bu be the restriction of u to $\partial\Omega$. Think of $B : X \to Y$, where Y is a space of functions on the boundary $\partial\Omega$. We are interested in cases where B is $\mathcal{A}$-closed, i.e. $f_n \in \mathcal{D}(\mathcal{A})$, $f_n \to f$, $\mathcal{A} f_n \to g$, $Bf_n \to h$ imply $f \in \mathcal{D}(\mathcal{A})$ and $Bf = h$. In our application, B is not closed but it is $\mathcal{A}$-closed.

Of concern is the problem (with general $\mathcal{A}, B$)

$$u'(t) = \mathcal{A}u(t) \quad (t \geq 0),$$

$$u(0) = f,$$

$$Bu(t) = g \quad (t \geq 0).$$

The compatibility condition at $t = 0$ is

$$Bf = g. \tag{10}$$

Express the solution as $u(t) = U(t; f, g)$.

We *assume* that this problem has a unique solution u and the function U satisfies

$$||U(t; f, g)|| \leq Me^{\omega t}(||f|| + ||g||)$$

for some $M \geq 1$, $\omega \in \mathbb{R}$ and all t, f, g such that (10) holds.

Now consider

$$u'(t) = \mathcal{A}u(t) + F(t) \quad (t \geq 0),$$
$$u(0) = f,$$
$$Bu(t) = w(t) \quad (t \geq 0).$$

We assume that A, the restriction of $\mathcal{A}$ to the null space $\mathbb{N}(B)$ of B, is the generator of a (C_o) semigroup T. (In our concrete example, $A = \Delta$ defined on $\mathcal{D}(A) = W^{2,p}(\Omega) \cap W_0^{1,p}(\Omega)$.)

Formally let

$$u(t) = U_1(t) + U_2(t)$$

where

$$U_1(t) = T(t)f + \int_0^t T(t-s)F(s)ds,$$
$$U_2(t) = U(t; 0, w(0) + \int_0^t U(t-s, 0, w'(s))ds.$$

Calculating gives

$$U_1' = \mathcal{A}U_1 + F, \qquad U_1(0) = f, \qquad BU_1 = 0,$$
$$U_2' = AU_2, \qquad U_2(0) = 0, \qquad Bu_2 = w.$$

Unfortunately this is not quite right because $U(t; f, g)$ is assumed to exist and have nice properties only when $Bf = g$.

We get around this by solving the abstract Dirichlet problem

$$\mathcal{A}u = 0, \quad Bu = g.$$

In our concrete example, this is $\Delta u = 0$ in Ω, $u = g$ on $\partial\Omega$. Call the solution $u = Gg$. Thus we assume there is a $G \in \mathcal{B}(Y, X)$ such that $\mathcal{A}Gg = 0$, $BGg = g$ for all $g \in Y$.

Now replace $\mathcal{A}$ by $\mathcal{A} - \lambda I$ for suitable real λ. Then

$$v = \lambda \int_0^\infty e^{-\lambda t} U(t; f, g) dt - \lambda(\lambda I - A)^{-1} f$$

satisfies $(\lambda I - \mathcal{A})v = 0$, $Bv = g$. Thus our incorrect formula $u = U_1 + U_2$ for the solution can be replaced by the correct formula

$$\begin{aligned} u(t) = T(t)(f - Gw(0)) + U(t; Gw(0), w(0)) \\ + \int_0^t T(t-s)(F(s) - Gw'(s))ds \\ + \int_0^t U(t-s; Gw'(s), w'(s))ds. \end{aligned}$$

One can verify that u is a solution in $C^1(\mathbb{R}^+, X)$ provided $w \in C^1(\mathbb{R}^+, Y)$, $f - Gw(0) \in \mathcal{D}(A)$, and $F - Gw' \in C^1(\mathbb{R}^+, X) + C(\mathbb{R}^+, Z)$ where $Z = \mathcal{D}(A)$, equipped with the graph norm. The formula for u is a kind of variation of parameters formula (or Duhamel formula) involving inhomogeneous "forcing terms" and inhomogeneous boundary conditions.

This leads to the solution of semilinear problems in the expected way. So consider

$$\begin{aligned} u'(t) &= Au(t) + F(t, u(t)), \\ u(0) &= f, \\ Bu(t) &= N(u(t)), \end{aligned}$$

where F and N are nonlinear but locally Lipschitzian (on suitable spaces). The successive approximation approach gives

$$\begin{aligned} u_n'(t) &= Au_n(t) + h_n(t), \\ u_n(0) &= f, \\ Bu_n(t) &= w_n(t), \end{aligned}$$

where

$$h_n(t) = F(t, u_{n-1}(t)),\ w_n(t) = N(u_{n-1}(t)).$$

Solve this and use the Banach fixed point theorem to get a local (in time) solution of the semilinear problem. Global existence follows in situations where one can establish an a priori bound.

MEAN ERGODIC THEOREM.

Let A generate a (C_o) contraction semigroup T on a Hilbert space $\mathcal{H}$. Then

$$\lim_{\tau\to\infty} \frac{1}{\tau}\int_0^\tau T(t)f dt = P_0 f.$$

Here P_0 is the orthogonal projections of $\mathcal{H}$ onto the null space of A, $\mathbb{N}(A) = \mathbb{R}(A)^\perp$. For the proof, let $F = f_1 + f_2 \in \mathbb{N}(A) + \mathbb{R}(A)$. Then $Af_1 = 0$, so $T(\cdot)f_1$ and f_1 are both solutions of $u' = Au$, $u(0) = f_1$; uniqueness implies $T(t)f_1 = f_1$ for all $t \geq 0$. Next,

$$\tau^{-1}\int_0^\tau T(t)f_2 dt = \tau^{-1}\int_0^\tau T(t)Agdt =$$

$$\tau^1\int_0^\tau \frac{d}{dt}(T(t)g)dt = \frac{T(\tau)g - g}{\tau} = o(1)$$

as $\tau \to \infty$. It follows that

$$\frac{1}{\tau}\int_0^\tau T(t)(f_1 + f_2)dt \to f_1$$

for $f_1 + f_2 \in \mathbb{N}(A) + \overline{\mathbb{R}(A)}$. In particular,

$$\mathbb{N}(A) \cap \overline{\mathbb{R}(A)} = \{0\}.$$

In the above arguments T and A can be replaced by $T^* = \{T(t)^* : t \geq 0\}$ and A^* respectively. Thus $\mathbb{N}(A^*) \cap \overline{\mathbb{R}(A^*)} = \{0\}$. But

$$\mathcal{H} = \{0\}^\perp = (\mathbb{N}(A^*) \cap \overline{\mathbb{R}(A^*)})^\perp = \overline{\mathbb{R}(A)} \oplus \mathbb{N}(A).$$

Since $||\frac{1}{\tau}\int_0^\tau T(t)fdt|| \leq ||f||$, thus $||P|| \leq 1$ and the result follows.

□

Let $\mathcal{HS}(\mathcal{H})$ be the set of all *Hilbert-Schmidt operators* on $\mathcal{H}$. For B a *compact* operator on $\mathcal{H}$, let $|B| = (B^*B)^{1/2}$; by the spectral theorem,

$$|B| = \sum_{n=1}^{\infty} \lambda_n < \cdot, e_n > e_n$$

where $\{e_n : n \in N\}$ is an orthonormal basis for $\mathcal{H}$. (Here we are taking $\mathcal{H}$ to be separable, although this is really not necessary.) The Hilbert-Schmidt norm of B is defined by

$$|||B|||^2 = ||\{\lambda_n\}||_{\ell^2}^2 = \sum_{n=1}^{\infty} \lambda_n^2,$$

and the inner product on the Hilbert space $\mathcal{HS}(\mathcal{H})$ is given by

$$\langle\langle B_1, B_2 \rangle\rangle = \text{ trace } (B_2^* B_1).$$

Let A be as in the mean ergodic theorem. Define $U = \{U(t) : t \in \mathbb{R}^+\}$ by

$$U(t)B = T(t)^* BT(t),$$

for $B \in \mathcal{HS}(\mathcal{H})$. Then U is a (C_o) contraction semigroup, and its generator G is given by

$$G(B) = A^*B + BA;$$

$G(B) = [B, A]$ holds when T is a unitary group. For f_j a unit vector in $\mathcal{H}$, let $B_j = \langle \cdot, f_j \rangle f_j = f_j \otimes f_j$. Then

$$\langle\langle U(t)B_1, B_2 \rangle\rangle = |\langle T(t)f_1, f_2 \rangle|^2$$

as a short calculation shows. The mean ergodic theorem implies

$$\frac{1}{\tau}\int_0^\tau |\langle T(t)f_1, f_2\rangle|^2 dt \to \langle\langle P_{N(G)}B_1, B_2\rangle\rangle,$$

and $G(B) = 0$ for $B \in \mathcal{HS}(\mathcal{H})$ implies B is a linear combination (or infinite series) of finite rank projections corresponding to purely imaginary eigenvalues of A. Thus we deduce

$$\frac{1}{\tau}\int_0^\tau |\langle T(t)f_1, f_2\rangle|^2 dt \to \sum_{\lambda\in\sigma_p(A)\cap i\mathbb{R}} |\langle P_\lambda f_1, f_2\rangle|^2.$$

This is *Wiener's theorem.* (Intuitively for the unitary part of T, $G(B) = 0 = AB - BA$, and the only compact operators which commute with $A^* = -A$ are built from rank one projections onto eigenvectors of A. For the nonunitary part of T, $|\langle T(t)f_1, f_2\rangle|^2$ formally tends to zero as $t \to \infty$.)

For a specific example, let μ be a finite Borel measure on $\mathbb{R}$, and let $\mathcal{H} = L^2(\mathbb{R}, \mu)$. Let A be multiplication by i times the identity function on $\mathcal{H}$. Then $(T(t)f)(x) = e^{itx}f(x)$ for $f \in \mathcal{H}$, x and $t \in \mathbb{R}$; and

$$\langle T(t)1, 1\rangle = \int_{-\infty}^{\infty} e^{itx}\mu(dx) = \widehat{\mu}(t),$$

is the (nonnormalized) Fourier transform of μ; and

$$\frac{1}{\tau}\int_0^\tau |\widehat{\mu}(t)|^2 dt \to \ell$$

by Wiener's theorem, where $\ell = \Sigma\{\langle P_\lambda 1, 1\rangle\}^2$. Unraveling what this means yields that ℓ is the sum of the squares of the jumps associated with μ. That is, if $\mu = \mu_c + \mu_d$ is the Lebesgue decomposition of μ and if the discrete part is given by $\mu_d = \Sigma a_j \delta_{x_j}$, then $\ell = \Sigma a_j^2$.

Now let L be a bounded linear operator on $\mathcal{H}$ such that $L(\lambda_0 - A)^{-1}$ is compact for some λ in the resolvent set of A. Then Wiener's theorem can be used to prove

$$\frac{1}{T}\int_0^T ||LT(t)f||^2 dt \to \sum_{\lambda\in\sigma_P(A)\cap i\mathbb{R}} ||LP_\lambda f||^2. \qquad (11)$$

This has the following interpretation when $A = -i(-\Delta + V(x))$ and $L = M_{\chi_{B(0,R)}}$, i.e. iA is a selfadjoint Schrödinger operator on $\mathcal{D}(A) = H^2(\mathbb{R}) \subset \mathcal{H} = L^2(\mathbb{R}^\ell)$ and L is multiplication by the characteristic function of the ball of radius R. (See the earlier section on the Feynman path formula.) Let u satisfy $u_t = Au, u(0) = f$ where f is a unit vector in $\mathcal{H}$. Then $|u(x,t)|^2$ is the position probability density of the system, i.e. $\int_\Gamma |u(x,t)|^2 dx$ is the probability that the position vector of the system lies in $\Gamma \subset \mathbb{R}^\ell$ at time t. We have

$$\frac{1}{T}\int_0^t ||LT(t)f||^2 dt = \frac{1}{T}\int_0^T \int_{|x|<R} |u(x,t)|^2 dx dt,$$

and

$$\lim_{R\to\infty}\lim_{T\to\infty}\frac{1}{T}\int_0^T \int_{|x|<R} |u(x,t)|^2 dx dt = 1\,[\text{ resp. } = 0] \qquad (12)$$

for *bound states* [resp. for *scattered states,* or for states that are asympotically free, and leave each bounded set as time gets large]. By (11), the above limit is

$$\lim_{R\to\infty}\sum_{\lambda\in\sigma_P(A)} ||P_\lambda f||^2 = 1 \quad [\text{ resp. } = 0] \qquad (13)$$

when f is in the closed span $\mathcal{H}_d(iA)$ of eigenvectors of A [resp. when f is orthogonal to this space]. For $H = H^* = \int_{-\infty}^{\infty} \lambda dE(\lambda)$ a self adjoint operator on a Hilbert space $\mathcal{H}$, we say that f is in the subspace $\mathcal{H}_{ac}(H)$ of absolute continuity of $\mathcal{H}$ iff the monotone function $\lambda \to ||E(\lambda)f||^2$ is absolutely continuous on $\mathbb{R}$. Similarly for $\mathcal{H}_{sc}(H), \mathcal{H}_c(H), \mathcal{H}_d(H)$, where sc [resp. c; resp. d] means singular continuous [resp. continuous; resp. discrete]. A monotone function is discrete when it is constant except for jumps. The Lebsegue decomposition theorem for selfadjoint operators says

$$\mathcal{H}_c(H) = \mathcal{H}_{ac}(H) \oplus \mathcal{H}_{sc}(H),$$

$$\mathcal{H} = \mathcal{H}_c(H) \oplus \mathcal{H}_d(H),$$

$$\mathcal{H}_d(H) = \overline{\text{span}}\{ \text{ all eigenvectors of } H\}.$$

What was shown in (12) and (13) together with the Lebesgue decomposition theorem implies that the scattering states [resp. bound states] are the vectors in $\mathcal{H}_c(H)$ [resp. $\mathcal{H}_d(H)$] for $H = iA$. In many applications, $\mathcal{H}_{sc}(H) = \{0\}$, and the scattering states are $\mathcal{H}_{ac}(H)$.

POSITIVITY AND SECOND ORDER OPERATORS.

If A is a (weakly) elliptic (partial) differential operator of order two, then by the maximum principle, the semigroup T generated by A is positivity preserving. So is the translation semigroup generated by the first derivative. A sort of converse holds; we indicate this is only the simplest situation.

Consider the lattice of real functions $L^2(\Omega, \mu.)$ Suppose a (C_o) semigroup T is given by

$$(T(t)f)(x) = \int_\Omega K(t,x,y)f(y)\mu(dy)$$

for all t, f, x. Then $K \geq 0$ a.e. on $\mathbb{R}^+ \times \Omega \times \Omega$ and $f \geq 0$ implies $T(t)f \geq 0$, i.e. T is a positivity preserving semigroup. From

$$(\lambda - A)^{-1}f = \int^\infty e^{-\lambda t}T(t)fdt,$$

$$T(t)f = \lim_{n\to\infty}(I - \frac{t}{n}A)^{-n}f,$$

it follows that $T(t)$ is positivity preserving for all $t \geq 0$ iff $(\lambda - A)^{-1}$ is positivity preserving for large $\lambda > 0$. But we want to know criteria for this involving A directly, not $(\lambda - A)^{-1}$.

Let $A = A^*$ be a nonpositive selfadjoint operator on $\mathcal{H} = L^2(\Omega, \mu)$, a real Hilbert space which we view as a Banach lattice of real functions. Let the (C_o) semigroup T generated by A be *positive* or *positivity preserving* in the sense that $f \in \mathcal{H},\quad f \geq 0$ implies $T(t)f \geq 0$. Then, since $f = f_+ - f_-$,

$$|T(t)f|^2 = \max\{|T(t)f_+|^2, |T(t)f|^2\} \leq |T(t)f_+|^2 + |T(t)f_-|^2$$

and consequently , by intergrating over Ω,

$$g(t) \equiv ||T(t)f||^2 \leq ||T(t)f_+||^2 + ||T(t)f_-||^2 \equiv h(t).$$

For $f \in \mathcal{D}(A)$, from $g \leq h$ on $[0, \infty]$ and $g(0) = h(0)$ we deduce $g'(0) \leq h'(0)$, i.e.,

$$\begin{aligned} \frac{d}{dt}\langle T(t)f, T(t)f\rangle|_{t=0} &= \langle Af, f\rangle = -||(-A)^{1/2}f||^2 \\ &\leq -||(-A)^{1/2}f_+||^2 - ||(-A)^{1/2}f_-||^2. \end{aligned}$$

Hence

$$||(-A)^{1/2}f_+||^2 + ||(-A)^{1/2}f_-||^2 \leq ||(-A)^{1/2}f||^2 < \infty,$$

and so $f \in \mathcal{D}(A)$ [or $f \in \mathcal{D}((-A)^{1/2})$] implies $f_+, f_- \in \mathcal{D}((-A)^{1/2})$, and the same holds for $|f|$. Suppose A is an elliptic operator of order 2m, so that

$$H_0^{2m}(\Omega) \subset \mathcal{D}(A) \subset H^{2m}(\Omega)$$

where these are the usual Sobolev spaces. Then $\mathcal{D}((-A)^{1/2}) \subset H^m(\Omega)$, and so $f \in H_0^{2m}(\Omega)$ implies $|f| \in H^m(\Omega)$. This implies $m \leq 1$. The simplest way to see this is to think about $f(x) \equiv$

x near $x = 0$; $f \in H^m(-1,1)$ for all m, but $|f| \in H^m(-1,1)$ holds only for $m \leq 1$.

SCATTERING THEORY.

For $j = 0, 1$, let H_j be a selfadjoint operator on a Hilbert space $\mathcal{H}$ and let U_j be the corresponding (C_o) unitary group generated by $-iH_j$; thus $U_j(t) = e^{-iH_j}$ governs the Schrödinger equation $idu/dt = H_j u$. The *free equation* [resp. *perturbed equation*] corresponds to $j = 0$ [resp. $j = 1$]. $V = H_1 - H_0$ is the perturbation. Think of V as a scattering source; thus $u_1(t) = U_1(t)f$ looks like a free solution for t large and negative and $u_1(t)$ looks like a different free solution for t large and positive. More precisely there are vectors $f_\pm \in \mathcal{H}$ such that

$$\lim_{t \to \pm\infty} ||U_1(t)f - U_0(t)f_\pm|| = 0.$$

Thus $f = W_\pm f_\pm$ where

$$W_\pm g = W_\pm(H_1, H_0)g = \lim_{t \to \pm\infty} U_1(-t)U_0(t)P_0 g$$

are the *wave operators*. Here P_0 is the orthogonal projection onto $\mathcal{H}_{ac}(H_0)$; when $W_\pm(H_1, H_0)$ exists, it is an isometry from $\mathcal{H}_{ac}(H_0)$ into $\mathcal{H}_{ac}(H_1)$. $W_\pm$ are called *asymptotically complete* when they exist and their ranges are both $\mathcal{H}_{ac}(H_1)$. The *scattering operator* S maps f_- to f_+. Thus

$$f_+ = W_+^{-1} f = W_+^{-1} W_- f_-$$

and so $S = W_+^{-1} W_-$ is unitary on $\mathcal{H}_{ac}(H_0)$ when $W_\pm$ exist and are complete. By the chain rule, $W_\pm(H_2, H_0) = W_\pm(H_2, H_1)W_\pm(H_1, H_0)$, and so existence and completeness of $W_\pm(H_1, H_0)$ follows from existence of both $W_\pm(H_1, H_0)$ and $W_\pm(H_0, H_1)$.

A closed densely defined operator A is said to be *H-smooth* provided there is a constant C such that

$$\int_{-\infty}^{\infty} ||AU(t)f||^2 dt \leq C^2 ||f||^2$$

holds for all f in a dense set in $\mathcal{H}$, where H is selfadjoint and $U(t) = e^{-itH}$. This implies

$$\mathbb{R}(A^*) \subset \mathcal{H}_{ac}(H).$$

Proof: Write $H = \int_{-\infty}^{\infty} \lambda dE(\lambda)$. For $f \in \mathcal{D}(A^*)$,

$$\sigma(\lambda) \equiv \langle E(\lambda)A^*f, A^*f\rangle = ||AE(\lambda)g||^2$$

(where $g = A^*f$) is bounded and nondecreasing, and

$$\begin{aligned} \sqrt{2\pi}\widehat{\sigma}(\xi) &= \int_{-\infty}^{\infty} e^{-i\xi\lambda} d\langle E(\lambda)g, g\rangle \\ &= \langle U(\xi)g, g\rangle = \langle AU(\xi)g, f\rangle; \end{aligned}$$

$$2\pi \int_{-\infty}^{\infty} |\widehat{\sigma}(\xi)|^2 d\xi \leq ||f||^2 \int_{-\infty}^{\infty} ||AU(\xi)g||^2 d\xi < \infty$$

since A is H-smooth. Thus $\widehat{\sigma} \in L^2(\mathbb{R})$ and so the monotone function σ is absolutely continuous and its derivative is in $L^2(\mathbb{R})$, by uniqueness of Fourier transforms. Hence $g \in \mathcal{H}_{ac}(H)$. □

The notion of (Kato) smoothness can be localized. For Γ a Borel set in $\mathbb{R}$, A is *H-smooth on Γ* means $AE(\Gamma)$ is H-smooth where

$$E(\Gamma) = \int_{\Gamma} dE(\lambda).$$

THEOREM.

Let $H_1 = H_0 + A_1^* A_0$ *in the quadratic form sense [i.e.* $\mathcal{D}(A_j) \supset \mathcal{D}(H_j)$, $\langle H_1 f, g\rangle = \langle f, H_0 g\rangle + \langle A_1 f, A_0 g\rangle$ *for* $f \in$

$\mathcal{D}(H_1)$, $g \in \mathcal{D}(H_0)$]. Suppose A_j is H_j-smooth on Γ for $j = 0, 1$. Then

$$W_\pm(\Gamma)f = \lim_{t\to\pm\infty} U_1(-t)U_0(t)E^0(\Gamma)f,$$
$$\Omega_\pm(\Gamma)f = \lim_{t\to\pm\infty} U_0(-t)U_1(t)E^1(\Gamma)f$$

exist for all $f \in \mathcal{H}$, where $H_j = \int_{-\infty}^{\infty} \lambda dE_j(\lambda), E^j(\Gamma) = \int_\Gamma dE_j(\lambda)$, $j = 0, 1$. If $\Gamma_n = (a_n, b_n) \subset \mathbb{R}$ is a sequence of intervals with $\sigma(H_j)\setminus \cup_{n=1}^{\infty} \Gamma_n$ at most countable, and if, for $j = 0, 1$, A_j is H_j-smooth on Γ_n for each n, then the wave operators $W_\pm(H_1, H_0)$ exist and are asymptotically complete.

This can be applied to $H_0 = -\Delta$, $H_1 = -\Delta + M_V$ on $L^2(\mathbb{R}^n)$, where V is real and belongs to $L^p(\mathbb{R}^n) + L^\infty(\mathbb{R}^n)$ for some $p > n/2$ (with $p > 1$), and V is a short range potential, i.e., $|V(x)| \le$ constant $|x|^{-1-\varepsilon}$ as $|x| \to \infty$. The idea is to take $A_j = M_{W_j}$ where $W_0 = |V|^{1/2}$, $W_1 = (\text{ sign } V)|V|^{1/2}$. One appeals to Weyl's theorem that, since H_1 is a relatively compact perturbation of H_0, $\sigma_{ess}(H_1) = \sigma_{ess}(H_0) = [0, \infty)$. Also any eigenvalues of H_1 in $(0, \infty)$ are isolated. Thus the Γ_n are chosen so that each interval Γ_n is in $\mathbb{R}^+$, is a positive distance from $\sigma_p(H_1)$, and $\cup_n \Gamma n = (0, \infty)\setminus\sigma_p(H_1)$. This result was proved by Kato and Kuroda (using different methods.) The notion of smoothness is due to Kato. Verifying that potential perturbations of elliptic operators are smooth is usually based on work of Agmon. Extensions of the Kato-Kuroda theorem from two body problems to N body problems was done by Enss, Sigal, Soffer and others. The books of Reed and Simon give valuable additional information.

We indicate the proof the above italicized thorem in the case when $\Gamma = \mathbb{R}$.

Let $w(t) = U_1(-t)U_0(t)h$. It suffices to show $\lim_{t\to+\infty} w(t)$ exists. The same argument applies to $\lim_{t\to-\infty} w(t)$ and for $\widetilde{w}(t) = U_0(-t)U_1(t)h$ since the hypotheses are symmetric in the indices 0, 1. Completeness of the wave operators then follows from the chain rule. Suppose $h \in \mathcal{D}(H_0)$, $k \in \mathcal{D}(H_1)$. Then for $t > s$,

$$\begin{aligned}
\frac{d}{dt}\langle k, w(t)\rangle &= -i\{\langle H_1U_1(t)k, U_0(t)h\rangle \\
&\quad - \langle U_1(t)k,\ H_0U_0(t)h\rangle\} \\
&= -i\langle A_1U_1(t)k, A_0U_0(t)h\rangle, \\
|\langle k, w(t) - w(s)\rangle| &\le \int_s^t |\langle AU_1(\tau)k, A_0U_0(\tau)h\rangle| d\tau \\
&\le \left(\int_s^t ||AU_1(\tau)k||^2 d\tau\right)^{1/2} \\
&\quad \cdot \left(\int_s^t ||A_0U_0(\tau)h||^2 dt\right)^{1/2} \\
&= J_1 + J_2;
\end{aligned}$$

$J_1 \le C||k||$ and $J_2 \to 0$ as $s, t \to \infty$, since A_j is H_j-smooth for $j = 0, 1$. Taking the supremum over $||k|| \le 1$ gives that $w(t)$ is Cauchy as $t \to \infty$. □

EQUIPARTITION OF ENERGY.

Let A be an injective selfadjoint operator on a Hilbert space $\mathcal{H}$. Of concern is the *abstract wave equation*

$$u''(t) + Au(t) = 0 \qquad (t \in \mathbb{R}),$$

$$u(0) = f_1, \quad u'(0) = f_2.$$

Define the *kinetic energy, potential energy,* and *total energy* at time t to be

$$K(t) = ||u'(t)||^2, \quad P(t) = ||Au(t)||^2, \quad E = K + P.$$

Think of $K(t)$ as the sum (or integral) over the space variable of the square of the velocity (times the constant density over two); thus $K(t)$ is the usual kinetic energy for $u_{tt} = c^2 u_{xx}$ for x in an interval J (say with Dirichlet boundary conditions) and $K(t) = \int_J |u_t(t,x)|^2 dx$.

The total energy is conserved. This follows from

$$\begin{aligned} E'(t) &= \frac{d}{dt}\{\langle u'(t), u'(t)\rangle + \langle Au(t), \quad Au(t)\rangle\} \\ &= 2Re\{\langle u'(t), u''(t)\rangle + \langle u'(t), A^2 u(t)\rangle\} \equiv 0 \end{aligned}$$

since $A^* = A$ and since $u'' + A^2 u = 0$. By the spectral theorem, we can write the unique solution of the abstract wave equation as

$$u(t) = \cos(tA)f_1 + A^{-1} \sin{(tA)} f_2.$$

(Note that for A selfadjoint, $A^{-1} \sin tA$ is a well-defined bounded linear operator, even when A fails to be injective, since $(\sin tx)/x \to t$ as $x \to 0$. Since

$$\begin{aligned} 2u'(t) &= e^{itA} g_1 + e^{itA} g_2, \\ 2iAu(t) &= e^{itA} g_1 - e^{-itA} g_2, \end{aligned}$$

with $g_1 = iAf_1 + f_2, g_2 = -iAf_1 + f_2$, it follows that

$$\begin{aligned} K(t) &= ||u'(t)||^2 = ||g_1||^2 + ||g_2||^2 + 2Re\langle e^{2itA} g_1, g_2\rangle, \\ P(t) &= ||Au(t)||^2 = ||g_1||^2 + ||g_2||^2 - 2Re\langle e^{2itA} g_1, g_2\rangle. \end{aligned}$$

Since $K + P = E$, and by the parrallelogram law,

$$E = E(0) = 2(||g_1||^2 + ||g_2||^2),$$

we conclude that as $t \to \pm\infty$,

$$K(t), P(t) \to E/2 \tag{14}$$

for all solutions if and only if $e^{itA} \to 0$ in the weak operator topology iff (for $A = \int_{-\infty}^{\infty} \lambda dE(\lambda)$)

$$\int_{-\infty}^{\infty} e^{it\lambda} d(||E(\lambda)h||^2) \to 0 \tag{15}$$

for all $h \in \mathcal{H}$. (In the last condition, $\langle E(\lambda)h, h\rangle$ can be replaced by $\langle E(\lambda)h, \ell\rangle$ using the polarization identity.) The condition (14) is called *asympotic equipartition of energy.*

A selfadjoint operator A satisfying (15) is called a *Riemann-Lebesgue operator.* More generally, for $A = \int_{-\infty}^{\infty} \lambda dE(\lambda)$ selfadjoint on $\mathcal{H}$, let $\mathcal{H}_{RL}(A)$ be the set of all vectors $h \in \mathcal{H}$ for which (15) holds. Then it can be shown that $\mathcal{H}_{RL}(A)$ is a closed subspace of $\mathcal{H}$ and

$$\mathcal{H}_{ac}(A) \subset \mathcal{H}_{RL}(A) \subset \mathcal{H}_c(A);$$

and each containment can be strict here. Since Δ on $L^2(\mathbb{R}^n)$ (and $(-\Delta)^{1/2}$ as well) is spectrally absolutely continuous, it follows that the concrete wave equation

$$u_{tt} = \Delta u \quad (x \in \mathbb{R}^n, \quad t \in \mathbb{R})$$

admits asymptotic equipartition of energy.

An extension of higher order problems was obtained by Goldstein and Sandefur. We shall indicate this in the next section.

ABSTRACT d'ALEMBERT FORMULA.

Let A_j generate a (C_o) semigroup $T_j = \{T_j(t) : t \in \mathbb{R}^+\}$ on X for $j = 1, \ldots, N$. We assume that the T_j are mutually commuting families:

$$T_j(t)T_k(s) = T_k(s)T_j(t)$$

holds for all j, k, t, s. Of concern is the equation

$$\prod_{j=1}^{N} \left(\frac{d}{dt} - A_j\right) u(t) = 0 \qquad (t \in \mathbb{R}^+). \tag{16}$$

We say that *d'Alembert's* formula holds if all solutions are of the form

$$u(t) = \sum_{j=1}^{N} T_j(t) f_j.$$

The classical example is

$$u_{tt} - u_{xx} = \left(\frac{d}{dt} - \frac{d}{dx}\right)\left(\frac{d}{dt} + \frac{d}{dx}\right) u = 0$$

with $X = L^2(\mathbb{R})$. Then every solution (in $L^2(\mathbb{R})$) is given by

$$\begin{aligned} u(t,x) &= f_1(x+t) + f_2(x-t) \\ &= \left(e^{t(+d/dx)} f_1\right)(x) \\ &+ \left(e^{t(-d/dx)} f_2\right)(x). \end{aligned}$$

Before generalizing this substantially, we note first that it does not always hold. In case $A_1 g = A_2 g$ for some vector $g \neq 0$, then

$$u(t) = tT_1(t)g$$

is a solution of (16) with $N = 2$.

We shall show informally why d'Alembert's formula holds whenever $A_j - A_k$ is injective and has sufficiently large range for $j \neq k$. (The above example $A_1 = -A_2 = d/dx$ on $L^2(\mathbb{R})$ indicates that we do not want to assume that $A_j - A_k$ is surjective.) We give the argument for $N = 2$ for convenience.

So consider $(d/dt - A_1)v(t) = 0$ where $v(t) = (d/dt - A_2)u(t)$. Then by the variation of parameters formula,

$$\begin{aligned} u(t) &= T_2(t)u(0) + \int_s^t T_2(t-s)v(s)ds \\ &= T_2(t)u(0) + \int_0^t T_2(t-s)T_1(s)v(0)ds \\ &= T_2(t)u(0) + \int_0^t \frac{d}{ds}[T_2(t-s)T_1(v)g]ds \\ &= T_2(t)[u(0)-g] + T_1(t)g \left(= \sum_{i=1}^{2} T_i(t)f_i\right) \end{aligned}$$

provided $g = (A_1 - A_2)^{-1} v(0)$. Thus $v(0) = u'(0) - A_2 u(0)$ should belong to the range of $A_1 - A_2$.

The abstract wave equation $u'' + A^2 u = 0$ (with $A = A^*$ on $\mathcal{H}$) can be written in factored from as

$$(d/dt - iA)(d/dt + iA)u(t) = 0.$$

Relative to $v = \binom{Au}{u_t}$, this becomes an abstract Schrödinger equation on $\mathcal{H}^2$ of the form

$$v_t = \mathcal{A}v, \qquad \mathcal{A} = \begin{pmatrix} 0 & A \\ -A & 0 \end{pmatrix}.$$

Let P_j be the orthogonal projection onto the jth component in $\mathcal{H}^2$. Then

$$K(t) = ||P_2 v(t)||^2, \ P(t) = ||P_1 v(t)||^2,$$

and as noted before, $K(t), P(t) \to E/2$ for all initial data (see (14)) if and only if A is Riemann-Lebesgue operator on $\mathcal{H}$ (in symbols, $A \in \mathbb{RL}(\mathcal{H})$), i.e. (15) holds, or $\mathcal{H}_{RL}(A) = \mathcal{H}$.

This formulation of equipartition of energy leads to a significantly deeper result due to Goldstein and Sandefur. Consider

$$\prod_{j=1}^{n} (d/dt - iA_j)u(t) = 0$$

where $A_1, \ldots, A_n$ are commuting selfadjoint operators on $\mathcal{H}$. The commuting hypothesis means that $\exp(i\,t\,A_j)$ and $\exp(i\,s\,A_k)$ commute for all t, s, j, k. When $n = 2^N$ is a power of 2, this equation can be written in the form

$$v' = i\,\mathcal{A}v.$$

where $\mathcal{A}$ is a selfadjoint operator on $\mathcal{K} = \mathcal{H}^n$. In this particular way of writing our factored equation as a system, let P_j be the orthogonal projection onto the jth component of $\mathcal{K}$; thus $\mathcal{K} = \bigoplus_{j=1}^{n} P_j\mathcal{K}$. Let $E_j(t) = ||P_j v(t)||^2$ be the jth *partial energy* of a solution u. Then the theorem is:

$E = \sum_{j=1}^{n} E_j(t)$ *is independent of* t; *and* $E_j(t) \to E/2^N$ *as* $t \to \pm\infty$ *for each* $j = 1, \ldots, 2^N$ *and all solutions if and only if* $A_j - A_k \in \mathrm{I\!R}\mathrm{I\!L}(\mathcal{H})$ *whenever* $j \neq k$.

The equations of linearized elasticity can be written in this form. Let u be any component of the shear wave or pressure wave. Then u satisfies

$$\sum_{j=1}^{n}(d^2/dt^2 - \lambda_j\Delta)u(t) = 0$$

where Δ is the Laplacian on $\mathcal{H} = L^2(\mathrm{I\!R}^3)$ and λ_1, λ_2 are the (positive) Lamé parameters. This can be written as a factored equation of order 4, and equipartition of energy holds if and only if $\lambda_1 \neq \lambda_2$.

Now consider the *telegraph equation* (or *damped wave equation*)

$$u'' + bu' + A^2u = 0, \tag{17}$$

where $A = A^*$ on $\mathcal{H}$ and b is a positive constant. If $K(t) = ||u'(t)||^2$, $P(t) = ||Au(t)||^2$ as before, then $K(t)$, $P(t) \to 0$ as $t \to \infty$, but one can still have equipartition of energy if K and P tend to zero at the same rate. The d'Alembert formula leads to the following result. If A is absolutely continuous (i.e. $\mathcal{H} = \mathcal{H}_{ac}(A)$), $A^2 \geq \alpha^2 I$, and if $0 < b < \alpha/2$, then for all nonzero solutions of (17),

$$\frac{K(t)}{P(t)} \to 1 \text{ as } t \to \infty.$$

SCATTERING REVISTED.

Consider the factored equation

$$\prod_{j=1}^{n} \left(\frac{d}{dt} + iH_j^{(k)} \right) u(t) = 0 \tag{$(17)_k$}$$

for $k = 0, 1$ and $t \in \mathbb{R}$. Here $H_j^{(k)}$ is a selfadjoint operator on $\mathcal{H}$. If $U_j^{(k)}(t) = \exp(-itH_j^{(k)})$, then the d'Alembert formula says that in many circumstances the solutions of $(17)_k$ are all given by

$$u(t) = \sum_{j=1}^{n} U_j^{(k)}(t) f_j.$$

Suppose that $\mathcal{K}_k = \mathcal{H}_{ac}\left(H_j^{(k)}\right)$ is independent of j. Suppose that, for each j, the wave operators

$$W_{\pm}\left(H_j^{(1)}, H_j^{(0)}\right)$$

exist and are asymptotically complete. This enables one to show that for each solution v of $(17)_1$, there are solutions $u_{\pm}$ of $(17)_o$ such that

$$||v(t) - u_{\pm}(t)|| \to 0$$

as $t \to \infty$. Using this one can construct wave and scattering operators for $(17)_k$. With this observation, together with the Birman-Kato invariance principle, one can do scattering theory for wave equations with potentials and for elasticity equations with potentials as a consequence of the known potential scattering theory for Schrödinger operators.

REMARKS ON NON (C_o) ***SEMIGROUPS.***

The solution of the heat equation on $\mathbb{R}^n$ is given by

$$u(t,x) = (T(t)f)(x) = (4\pi t)^{-n/2} \int_{\mathbb{R}^n} e^{-|x-y|^2/4t} f(y)dy.$$

T defined in this way defines a (C_o) semigroup on various spaces, e.g., $L^p(\mathbb{R}^n)$ $(1 \le p < \infty)$, $C_o(\mathbb{R}^n)$, $C(\overline{\mathbb{R}^n})$, and $BUC(\mathbb{R}^n)$. But this formula actually defines a semigroup on larger spaces, for instance $X = L^\infty(\mathbb{R}^n)$. But T on X is not strongly continuous at $t = 0$. It satisfies $T(t) : X \to BUC(\mathbb{R}^n)$ for $t > 0$, and the largest subspace of X on which T is strongly continuous on $\mathbb{R}^+$ is $BUC(\mathbb{R}^n)$.

Now consider

$$u_t = \Delta u + h(t),\ u(0) = f$$

where Δ is the Dirichlet Laplacian acting on real or complex functions on Ω, a smooth bounded set in $\mathbb{R}^n$. Δ generates a (C_o) semigroup on $Y = C_o(\Omega)$, but it determines a semigroup T on $X = C(\bar{\Omega})$, which is not strongly continuous at $t = 0$. The variation of parameters formula gives

$$u(t) = T(t)f + \int_0^t T(t-s)h(s)ds. \tag{18}$$

This is classical for $h : \mathbb{R}^+ \to Y$, but in this case $h(t,x)$ should vanish for $x \epsilon \partial\Omega$. This is not always reasonable, and one wants often to use (18) for $h : \mathbb{R}^+ \to X$.

Here is another example of a non (C_o) semigroup. Let $X = C_o[0,1) = \{f \in C[0,1] : f(1) = 0\}$. One can view X as a subspace of $C_o(\mathbb{R}^+)$ by defining $f(x) = 0$ for $x > 1$ and $f \in X$. Let $A_1 = d/dx$ and A_2 be multiplication by $V \in L^1[0,1]$. The Daletskii-Lie-Trotter product formula gives that the semigroup T generated by $A = A_1 + A_2$ is given by

$$(T(t)f)(x) = \exp\{\int_0^t V(x+s)ds\} f(x+t),$$

since

$$\left[T_1(\frac{t}{n})T_2(\frac{t}{n})\right]^n f(x) = \exp\{\sum_{j=1}^n \frac{t}{n} V(x + jt/n)\} f(x+t).$$

If we take $V(x) = c/x$, so that

$$\exp\{\int_0^t V(x+s)ds\} = \exp\{c \log\left(\frac{x+t}{x}\right)\} = \left(\frac{x+t}{x}\right)^c,$$

then we get a semigroup T given by

$$T(t)f(x) = \left(\frac{x+t}{x}\right)^c f(x+t)$$

with generator $A = \frac{d}{dx} + \frac{c}{x}$. For$-1 < c < 0$ and $t > 0$, $T(t)$ is not a bounded operator on X, but $S(t)$ is, where

$$S(t)f = \int_0^t T(s)f ds.$$

This T is also an integrated semigroup.

We now give one of several equivalent definitions of an n-times integrated semigroup. The case of $n = 0$ corresponds to a C_0 semigroup. A closed linear operator A is the generator of an *n–times integrated semigroup* on X (for $n = 0, 1, 2, \ldots$) if for all $f \in \mathcal{D}(A)$, there is a unique $w \in C(\mathbb{R}^+, X)$ such that $\int_0^t w(s)ds$ belongs to $\mathcal{D}(A)$ and

$$w(t) = A\int_0^t w(s)ds + \frac{t^n}{n!}f$$

holds on $\mathbb{R}^+$. If $w \in C^{n+1}(\mathbb{R}^+, X)$, then $u = w^{(n)}$ satisfies

$$u'(t) = Au(t), \ u(0) = f. \tag{19}$$

For $n = 0$, one simply interprets this to mean $u \epsilon C^1(\mathbb{R}^+, X)$ and (19) holds.

It is a theorem that an operator A is the generator of an n-times integrated semigroup iff the initital value problem

$$du/dt = Au, \ u(0) = f$$

has a unique solution $u \epsilon C^1(\mathbb{R}^+, X) \cap C(\mathbb{R}^+, \mathcal{D}(A))$ for all $f \in \mathcal{D}(A^{n+1})$ satisfying

$$||u(t)|| \leq Me^{wt}\sum_{j=0}^{n}||A^j f||$$

where $M \geq 1$ and $w \in \mathbb{R}$ are constants, independent of f. Let A generate an n-times integrated semigroup on X. Then, according to the theorem of Arendt, Neubrander, and Schlotterbeck, there are Banach spaces Y, Z with continuous dense embeddings

$$\mathcal{D}(A^{n+1}) \hookrightarrow Y \hookrightarrow X \hookrightarrow Z$$

so that the natural extension and restriction of A to Z and Y, $A|_Z$ and $A|_Y$, both generate (C_o) semigroups, on Z and Y respectively. Thus is some sense, integrated semigroup theory reduces to (C_0)

semigroup theory. But in other ways, integrated semigroups are indeed more general.

The theory of non (C_0) semigroups is a rapidly developing field; the theory is developing faster than the applications. Interesting applications and examples have been obtained by W. Arendt and by M. Hieber, and by O. Diekmann and his collaborators.

AFTERWORD.

We have made no effort to credit original sources or to give a comprehensive bibilography. Most of these notes follow material in my book [1]. For most of the rest of the material, see the papers [2]-[7] and the references contained therein.

The author gratefully acknowledges the partial support of the National Science Foundation.

The theory and practice of semigroups of linear operators remains an active subject. It seems certain that in 1997 one will be able to write a substantial survey paper on applications which will be based on work done since 1992.

REFERENCES

1. J. A. Goldstein, *Semigroups of Linear Operators and Applications,* Oxford U. Press, New York and Oxford, 1985.

2. J.A. Goldstein, Asymptotics for bounded semigroups on Hilbert space, in *Aspects of Positivity in Functional Analysis* (ed. by R. Nagel, U. Schlotterbeck, and M.P.H. Wolff), Elsevier (North-Holland), Dordrecht (1986), 49-62.

3. J.A. Goldstein, Evolution equations with nonlinear bound-

ary conditions, in *Nonlinear Semigroups, Partial Differential Equations and Attractors* (ed. by T. L. Gill and W. W. Zachary), Lecture Notes in Math. No. 1248, Springer, Berlin (1987), 78-84.

4. J. A. Goldstein (in collabration with W. Caspers, K. Engel, M. Hieber, F. Räbiger, and G. R. Rieder), Lectures on Applications of Semigroups of Linear Operators, *Semesterbericht Funktionalanalysis, Tübingen* (1990), 59-139.

5. J. A. Goldstein and J.T. Sandefur, Jr., Equipartition of energy for higher order abstract hyperbolic equations, *Comm. P.D.E.* 7 (1982), 1217-1251.

6. J. A. Goldstein and J.T. Sandefur, Jr., An abstract d'Alembert formula, *SIAM J. Math. Anal.* 18 (1987), 842-856.

7. J.A. Goldstein, R. de Laubenfels, and J. T. Sandefur Jr., Regularized semigroups, iterated Cauchy problems, and equipartition of energy, *Monatshefte Math.* (1993).

Nonlinear Semigroups and Applications

Gisèle Ruiz Goldstein

1. Introduction

Our aim is to study problems which are governed by the abstract Cauchy problem

$$\frac{du(t)}{dt} = A\big(u(t)\big) \qquad t > 0$$
$$u(0) = f. \tag{ACP}$$

Here X is a Banach space, $u(t)$ takes values in X for each $t > 0$, and A is an operator (not necessarily linear) which maps its domain $\mathcal{D}(A) \subseteq X$ into X. We are interested in questions of existence and uniqueness of solutions (in some sense) of (ACP), as well as continuous dependence of solutions on the initial data. A problem which satisfies these three conditions is said to be *well-posed.* We shall also be concerned with qualitative features of the solution, such as regularity.

Problems which can be formulated as an abstract Cauchy problem arise in wide variety of applications, including physics, chemistry, fluid mechanics, mathematical biology, probability theory, quantum theory, and differential geometry. We mention a few important examples which fit into this framework.

Example 1: Let Ω be an open subset of $\mathbb{R}^n, \bar{\Omega} \subsetneqq \mathbb{R}^n$, with "nice" boundary $\partial\Omega$. The heat equation in Ω can be written as the initial value problem

$$\begin{aligned} \frac{\partial v}{\partial t} &= \Delta v \qquad t > 0 \\ v(0,x) &= f(x) \qquad x \in \Omega. \end{aligned} \tag{1.1}$$

The problem (1.1) may be studied with a wide variety of boundary conditions; for definiteness we consider Dirichlet boundary

G. R. Goldstein and J. A. Goldstein (eds.),
Semigroups of Linear and Nonlinear Operations and Applications, 59–98.

conditions

$$v(t,x) = 0 \qquad\qquad t \geq 0,\ x \in \partial\Omega. \tag{1.2}$$

Let X be a Banach space consisting of functions on Ω, and let $A = \Delta$. If we define the domain of A by $\mathcal{D}(A) = \{w \in X \text{ and } w = 0 \text{ on } \partial\Omega\}$, then this problem is of the form (ACP). One can prove that this problem is "good"; that is, it has a unique solution which depends continuously on the initial data in many spaces X.

Example 2: We consider the initial value problem for the wave equation on $\mathbb{R}^n$.

$$\begin{cases} \frac{\partial^2 v}{\partial t^2} = \Delta v & \text{for } t > 0, x \in \mathbb{R}^n \\ v(0,x) = f(x) & \text{for } x \in \mathbb{R}^n \\ \frac{\partial v}{\partial t}(0,x) = g(x) & \text{for } x \in \mathbb{R}^n. \end{cases} \tag{1.3}$$

Set

$$u(t,x) = \begin{pmatrix} v(t,x) \\ v_t(t,x) \end{pmatrix}, F(x) = \begin{pmatrix} f(x) \\ g(x) \end{pmatrix}, A = \begin{pmatrix} 0 & I \\ \Delta & 0 \end{pmatrix},$$

and let $u(t,\cdot) \in X$, where X represents a space of pairs of functions on Ω. Then (1.3) formally reduces to the form (ACP). One can show that this is a "good" problem only in norms related to the energy norm, where

$$\left|\left| \begin{pmatrix} f_1 \\ f_2 \end{pmatrix} \right|\right|^2_{\text{energy}} = ||\nabla f_1||^2_{L^2(\mathbb{R}^n)} + ||f_2||^2_{L^2(\mathbb{R}^n)}.$$

Here the first and second terms on the right side represent the potential energy and the kinetic energy respectively.

In Examples 1 and 2 the operator A is linear. We now give some examples which illustrate the importance of nonlinear operators A as well.

Example 3: Consider the initial value problem for the one dimensional Hamilton-Jacobi equation

$$\begin{cases} \frac{\partial v}{\partial t} + H\left(\frac{\partial v}{\partial x}\right) = 0 & \text{for } t > 0 \text{ and } x \in \mathbb{R} \\ v(0,x) = f(x) & \text{for } x \in \mathbb{R}. \end{cases} \tag{1.4}$$

Let X be a space of functions on $\mathbb{R}$. Define the operator A by $A = -H \circ \frac{d}{dx}$. Clearly, A is nonlinear. This problem is "good" if we take X to be the space of bounded uniformly continuous functions on $\mathbb{R}$; it is not a nice problem in any L^p norm for $1 \leq p < \infty$.

Example 4: Flow through a porous medium in a region $\Omega \in \mathbb{R}^n$ can be modeled by the following initial-boundary value problem

$$\begin{cases} \frac{\partial v}{\partial t} = \Delta\big(\varphi(v)\big) & \text{for } t > 0, x \in \Omega \\ v(0,x) = f(x) & \text{for } x \in \Omega \\ v(t,x) = 0 & \text{for } x \in \partial\Omega. \end{cases} \tag{1.5}$$

Let $\varphi(v) = v^{m+1} = |v|^m v$ for $m > 0$, although one can allow for more general functions φ. As we shall see later, this problem can be put into the form (ACP); it is a "good" problem in either the space $L^1(\Omega)$ or $H^{-1}(\Omega)$.

Note that if $m = 0$, we have the linear heat equation. However, the solutions of (1.5) and (1.1), (1.2) have dramatically different features. In the case of the heat equation, with $\Omega = \mathbb{R}^n$, it is well known that the solution u may be written in the form

$$u(t,x) = (G_t * f)(x)$$

where G_t is the Gaussian heat kernel. Hence, if the initial data $f(x)$ is positive on some set of positive measure, the solution $u(t,x) > 0$ for all $x \in \mathbb{R}^n$ and all $t > 0$. We say the equation has an infinite speed of propagation. In a later section we shall show that the porous medium equation has a finite speed of propagation.

Suppose that our abstract Cauchy problem is well-posed. Let $\{T(t) : t \geq 0\}$ be a family of operators on X, and suppose that $T(t)$ maps the solution of (ACP) at time s to the solution at time $t + s$. The assumption that the operator A is independent of time is just the assumption that $T(t)$ does not depend on s. The solution at time $t + s$ can be represented by $T(t+s)f$ where $u(0) = f$. On the other hand, if we start the equation at $t = 0$, let the system run for s units of time, then use the solution at time s as new initial data and run the sytem for t units of time longer, we see that by uniqueness we must have

$$T(t+s)f = T(t)T(s)f.$$

In addition we require $T(0) = I$ and, $t \to T(t)f$ is differentiable for $t \geq 0$ and $\frac{d}{dt}T(t)f = AT(t)f$, so that $u(t) = T(t)f$ and $T(t)$ is continuous on X; hence the solution depends continuously on the initial data. Finally, the initial data should be in $\mathcal{D}(A)$. We assume $\overline{\mathcal{D}(A)} = X$. Note that $T(t)$ is linear iff A is linear.

Consider the case where A (or equivalently $T(t)$) is linear. The family $\{T(t) : t \geq 0\}$ of bounded linear operators from X into X is a *C_0-semigroup (or strongly continuous semigroup)* if

$$\begin{cases} T(t+s) = T(t)T(s) & \text{for all } t, s \in \mathbb{R}^+ \\ T(0) = I \\ T(\cdot)f \in C\big([0,\infty); X\big) & \text{for all } f \in X. \end{cases} \tag{1.6}$$

The family $\{T(t) : t \geq 0\} \subseteq B(X)$ is a *C_0-semigroup of contractions* if $||T(t)|| \leq 1$. In fact suppose $\{T(t) : t \geq 0\}$ is a C_0-semigroup. By the Uniform Boundedness Principle

$$L = \sup_{0 \leq t \leq 1} ||T(t)|| < \infty.$$

Let $n = [t]$; then $T(t) = T(t-n)T(n)$ and $||T(t)|| \leq L^{n+1} \leq Le^{\omega t}$ where $\omega = lnL$. Set

$$S(t) = e^{\omega t}T(t);$$

then $||S(t)|| \leq L$. If we define a new norm $|||\cdot|||$, on X by $|||f||| = \sup_{t>0} ||S(t)f||$, then

$$\begin{aligned} |||T(t)f||| &= \sup_{\sigma>0} ||S(\sigma)T(t)f|| = \sup_{\sigma>0} ||e^{-\omega\sigma}T(\sigma)T(t)f|| \\ &= \sup_{\tau>t} ||e^{\omega\sigma}T(\tau)f|| \\ &\leq \sup_{\tau>0} ||e^{-\omega(\tau-t)}T(\tau)f|| = e^{\omega t}|||f|||. \end{aligned}$$

Hence, $|||S(t)||| \leq 1$, and clearly $|||\cdot|||$ and $||\cdot||$ are equivalent norms. Roughly speaking, this calculation shows it is sufficient to consider only C_0-contraction semigroups.

The following results are the basis for the theory of semigroups of linear operators.

Theorem 1: *(Hille-Phillips) The abstract Cauchy problem (ACP) is "well-posed" iff it is governed by a C_0-semigroup $\{T(t) : t \geq 0\}$ iff A is the infinitesimal generator of a C_0-semigroup T. Here $Af = \lim_{t \downarrow 0} \frac{T(t)f - f}{t}$, and $f \in \mathcal{D}(A)$ iff this limit exists. In this case the unique solution of (ACP) is $u(t) = T(t)f$.*

Formally this suggests that $T(t) = e^{tA}$ where $A = T'(0)$.

Theorem 2: *(Hille-Yosida Generation Theorem) The operator A generates a contraction C_0-semigroup T iff $D(A)$ is dense in X and the following conditions hold:*

$$||(I - \lambda A)^{-1}|| \leq 1 \qquad \textit{for all } \lambda > 0 \tag{1.7}$$

$$R(I - \lambda A) = X. \tag{1.8}$$

The theory of semigroups has been extensively studied. Applications have been found in a wide range of areas. For more study of some aspects of the linear theory, the interested reader should refer to the preceding article by J. A. Goldstein as well as [18] [24] [28] [29]. One of the most amazing facts about the theory of semigroups of linear operators is that *linearity is irrelevant.* This outrageous statement is in some sense accurate. In the remainder of this paper we assume only that A is an operator on the Banach space X, where A is either linear or nonlinear.

Let us proceed formally. Consider the backward difference scheme for (ACP),

$$\frac{u_\varepsilon(t) - u_\varepsilon(t - \varepsilon)}{\varepsilon} = Au_\varepsilon(t). \tag{1.9}$$

Then (1.9) may be written as

$$\begin{cases} u_\varepsilon(t) = (I - \varepsilon A)^{-1} u_\varepsilon(t - \varepsilon) \\ u_\varepsilon(0) = f. \end{cases}$$

Letting $\varepsilon = \frac{t}{n}$, we see that

$$u_\varepsilon(t) = (I - \frac{t}{n}A)^{-n} f. \tag{1.10}$$

We are concerned with finding the limit of $u_\varepsilon(t)$ as $n \to \infty$. In the applications usually A is some sort of differential operator (and hence is unbounded). However, in that case the operators $(I-\varepsilon A)^{-1}$ are integral operators which are smoothing.

On the other hand, if instead of (1.9) we use the forward difference scheme

$$u_\varepsilon(t)-u_\varepsilon(t-\varepsilon)=A\left(u_\varepsilon(t-\varepsilon)\right), \tag{1.11}$$

then the analogue of (1.10) is

$$u_\varepsilon(t)=(I+\frac{t}{n}A)^{-n}f. \tag{1.12}$$

Bu the operators $(I+\frac{t}{n}A)^{-n}$ are differential operators; hence $u_\varepsilon(t)$ doesn't converge in general even in the linear case.

Our first goal is to find a nonlinear analogue of the Hille-Yosida theorem. We define the *Lipschitz seminorm* of an operator S on X by

$$||S||_{Lip}=\inf\{C\in\mathbb{R}^+:||Sf-Sg||\leq C||f-g|| \quad \text{for all} f,g\in X\}.$$

If $||(I-\lambda A)^{-1}||_{Lip}\leq 1$, we say A is *dissipative.* If A is dissipative and $R(I-\lambda A)=X$ for some (and hence all) $\lambda>0$, then A is *m-dissipative.* We define the semigroup $T=\{T(t):t\in\mathbb{R}^+\}$ generated by A via the formula

$$T(t)f=\lim_{n\to\infty}(I-\frac{t}{n}A)^{-n}f \tag{1.13}$$

for all $f\in\overline{\mathcal{D}(A)}$. T is a *contraction semigroup* if $||T(t)||_{Lip}\leq 1$ for all $t\geq 0$.

The following theorems form the nonlinear analogues of the Hille-Yosida and Hille-Phillips theorems.

***Theorem 3:**(Crandall-Liggett) If A is m-dissipative on X, then A generates a strongly continuous contraction semigroup on $\overline{\mathcal{D}(A)}$.*

***Theorem 4:** (Bénilan-Kobayashi) If A is m-dissipative, then (ACP) is well-posed.*

Note that in the nonlinear case the implications are only in one direction; it is, however, the most important direction for the applications.

Let us make a few comments about these theorems. The semigroup of Theorem 3 is given by

$$T(t)f = \lim_{n\to\infty}(I - \frac{t}{n}A)^{-n}f \qquad \text{for } f \in X,$$

and the solution of (ACP) is given by $u(t) = T(t)f$. In general u may be nowhere differentiable, but if X has the Radon-Nikodym property, then $u'(t) \in Au$ holds a.e. for $f \in \mathcal{D}(A)$. (Recall that all reflexive spaces have the Radon-Nikodym property). For a general Banach space X the notion of solution is in a very general sense, which we explain in the next section. Hence, the notion of well-posedness in Theorem 4 must be general enough to allow for this notion of solution.

Finally we note that the converse of Theorem 3 is false in general, but holds in the Hilbert space case.

This paper is based on the short course which I presented in Curaçao. I warmly and sincerely thank Jerry Goldstein for his lecture notes, which are soon to be a book [19] on this topic. Sections 1-5 of this paper follow closely Jerry's notes.

§2. The Notion of Solution-An Introduction

Let H be a Hilbert space, and let A be a *linear* operator on H. Then the condition (1.7) in the Hille-Yosida theorem is equivalent to

$$Re\langle Af, f\rangle \le 0 \qquad \text{for all } f \in \mathcal{D}(A). \tag{2.1}$$

The corresponding condition in the nonlinear analogue of the Hille-Yosida theorem is

$$\|(I - \lambda A)^{-1}\|_{Lip} \le 1. \tag{2.2}$$

(Here A may be nonlinear). If A is linear, clearly (2.2) reduces to (1.7). In general (2.2) is equivalent to

$$Re\langle Af - Ag, f - g\rangle \le 0 \qquad \text{for all } f, g \in \mathcal{D}(A). \tag{2.3}$$

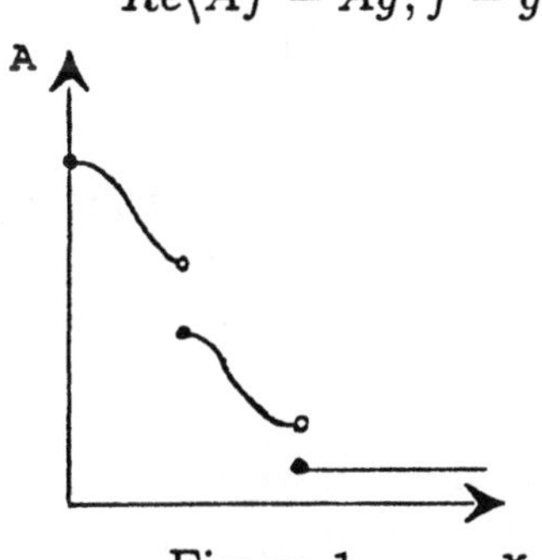

Figure 1.

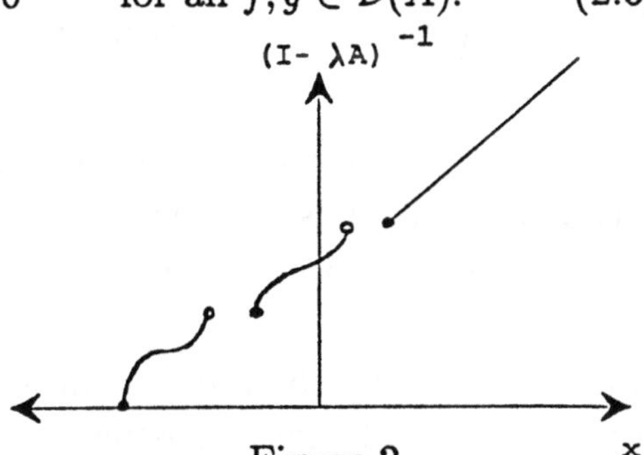

Figure 2.

Let us look at the example $A : \mathbb{R} \to \mathbb{R}$, where A is a single-valued function of one (real) variable. Clearly (2.3) holds iff A is a nonincreasing function. Since A is not necessarily continuous, the range of $(I - \lambda A)$ is not necessarily all of $\mathbb{R}$. (See Figures (1) and (2)).

In order to make $(I - \lambda A)^{-1}$ everywhere defined, we must "fill-in" the gaps, that is to say we must allow the function A to be multivalued. If A has a discontinuity at x_0, we redefine A so that its value at x_0 is the interval $\left[A(x_0+), A(x_0-)\right]$ where $A(x_0\pm) = \lim_{x \to x_0^{\pm}} A(x)$ Allowing A to be multivalued allows the range condition, $R(I-\lambda A) = X$, in the Crandall-Liggett Theorem to be satisfied.

We begin our discussion of the notion of solution for the nonlinear problem (ACP) by considering some examples. We begin with a nonlinear conservation law in one dimension

$$\begin{cases} \frac{\partial u}{\partial t} + \frac{\partial}{\partial x}(\varphi(u)) = 0 \\ \\ u(x,0) = f(x) \end{cases} \qquad x \in \mathbb{R}, t > 0 \tag{2.4}$$

where $\varphi : \mathbb{R} \to \mathbb{R}$ is a smooth function.

Suppose u is a classical solution of (2.4). Then on the characteristic curves $x(t)$, defined by

$$\frac{dx}{dt} = \varphi'\Big(u\big(x(t), t\big)\Big),$$

we have, by the equation,

$$\frac{d}{dt}\Big(u\big(x(t), t\big)\Big) = u_x \frac{dx}{dt} + u_t = u_x \varphi'(u) + u_t = 0.$$

Thus, u is constant along characteristics, and it follows that $\frac{dx}{dt}$ is constant. That is, characteristics of (2.4) are straight lines in the (x,t) plane. (See Figure 3.)

In order to satisfy the initial condition, we must have $u(x_0, t_0) = f(x^*)$. In general characteristics can and do intersect, so we can't have classical solutions to (2.4). (See Figure 4.)

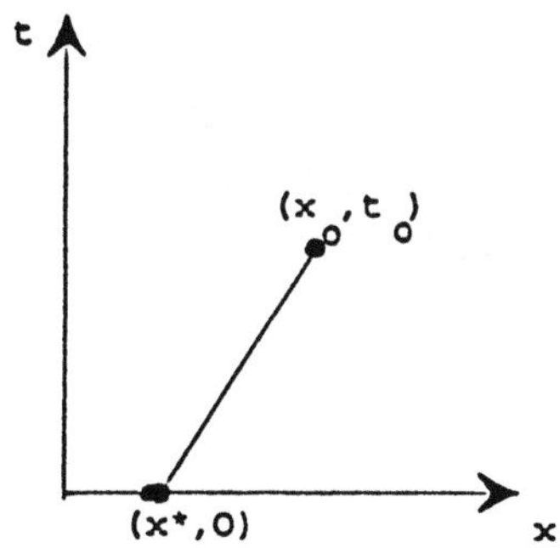

Figure 3

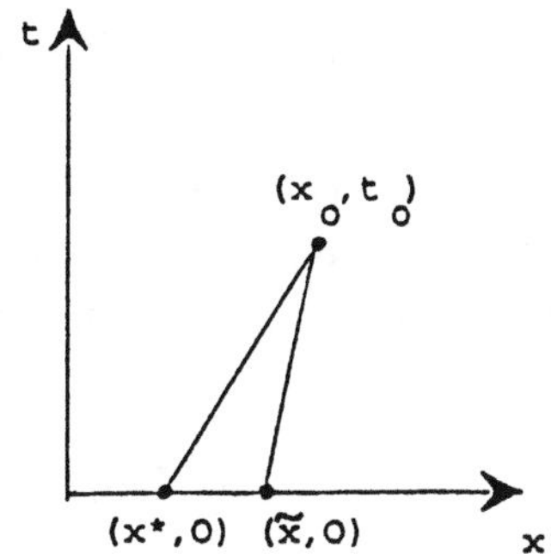

Figure 4

For example in one space dimension if we consider the Hopf equation $u_t + uu_x = 0$ with $u(x_0, t_0) = f(x)$, then the characteristic equation is

$$\frac{dx}{dt} = u(x,t).$$

Hence $\varphi'(u) = u$, so $\varphi(u) = \frac{u^2}{2}$; in particular φ is convex. Let x^* and $\tilde{x}$ be as in Figure 4. Clearly the characteristics through $(x^*, 0)$ and $(\tilde{x}, 0)$ intersect whenever $f(\tilde{x}) < f(x^*)$. If we want to allow initial data with compact support, then this example shows we cannot expect classical solutions. Hence, if we want global solutions we must weaken our notion of solution to allow for discontinuous solutions.

We can define solutions to (2.4) by multiplying the equation by $g \in C^\infty(\mathbb{R})$ and integrating by parts. A measurable function $u \in L^\infty_{loc}(\mathbb{R})$ is a *weak solution* of (2.4) if

$$\int_{-\infty}^{\infty}\int_0^{\infty} [-g_t u - g_x \varphi(u)]\, dt dx + \int_{-\infty}^{\infty} f(x) g(x,0) dx = 0. \quad (2.5)$$

This condition is sufficiently weak to allow for the existence of global solutions. However, solutions defined by (2.5) are not unique. For example if we again consider the Hopf equation we can easily verify that

$$u_1(x,t) = \begin{cases} 0 & x < \frac{t}{2} \\ 1 & x \geq \frac{t}{2} \end{cases} \qquad u_2(t) = \begin{cases} 0 & x \leq 0 \\ \frac{x}{t} & 0 < x < t \\ 1 & x \geq t \end{cases}$$

are both weak solutions, with the same initial data; in fact there are infinitely many weak solutions with this data. Somehow we must find a way to single out the physically correct solution.

Oleinik's idea was first to regularize the problem (2.4), that is, find u^ε (for ε small) which satisfies

$$\begin{cases} u_t^\varepsilon + \varphi(u^\varepsilon)_x &= \varepsilon u_{xx}^\varepsilon \\ u^\varepsilon(0,x) &= f(x). \end{cases} \tag{2.6}$$

Set $u(t,x) = \lim_{\varepsilon \to 0+} u^\varepsilon(t,x)$. Then this solution exists; this method is known as the method of vanishing viscocity. We shall require a further condition, known as an "entropy condition", to guarantee uniqueness of a generalized solution.

Let $g \in C_c^\infty(\mathbb{R} \times [0,\infty))$ and $\Phi \in C^2(\mathbb{R})$. It is easy to see that

$$\Phi(u) = \Phi'(u)u_r \qquad \text{for } r = x,t,$$

and for $k \in \mathbb{R}$,

$$\varphi'(u)u_x\Phi(u) = \left[\int_k^u \varphi'(s)\Phi'(s)ds\right]_x.$$

Multiplying the equation (2.6) by $g\Phi(u)$ and integrating by parts yields

$$\begin{aligned} I := &\int_0^\infty \int_{-\infty}^\infty \left[\Phi(u^\varepsilon)g_t + g_x\left(\int_k^{u^\varepsilon} \varphi'(s)\Phi'(s)ds\right)\right] dxdt \\ = \varepsilon &\int_0^\infty \int_{-\infty}^\infty \left[u_x^\varepsilon g_x \Phi'(u) + (u_x^\varepsilon)^2 g\Phi''(u^\varepsilon)\right] dxdt \\ = -\varepsilon &\int_0^\infty \int_{-\infty}^\infty \Phi(u^\varepsilon)g_{xx} + \varepsilon \int_0^\infty \int_{-\infty}^\infty (u_x^\varepsilon)^2\, g\Phi''(u^\varepsilon). \end{aligned}$$

The last term on the right side involves $(u_x^\varepsilon)^2$ which is difficult to estimate. However, if we assume g is nonnegative and Φ is convex, then this term is nonnegative. Thus,

$$I \geq -\varepsilon \int_0^\infty \int_{-\infty}^\infty \Phi(u^\varepsilon)g_{xx},$$

and letting $\varepsilon \downarrow 0$, we see that $I \geq 0$.

Actually, it is sufficient to consider the function $\Phi(s) = |s-k|$. This method, developed by Kruzkov, gives a notion of solution to (2.4) which is defined by a family of inequalities. This solution ($u = \lim u^{\varepsilon}$) exists, is unique and may be discontinuous. For our example of a single conservation law, a solution defined in this way allows for downward jumps only, provided φ is convex.

Notice that if v is a classical solution of the Hamilton-Jacobi equation (1.4) and if we set $u = \frac{\partial v}{\partial x}$, the Hamilton-Jacobi equation reduces to (2.4) with $H = \varphi$. Solutions of the Hamilton-Jacobi equation may therefore be expected to be continuous, but their spatial derivatives may have discontinuities.

The physically correct space in which to consider the Hamilton- Jacobi equation is a space involving the supremum norm, while for the conservation law the correct space is $L^1(\mathbb{R})$. Hence we must allow the theory of nonlinear semigroups to be general enough to include the notion of solution indicated above, and we must allow for nonreflexive Banach spaces.

3. A sketch of the proof of the Crandall-Liggett Theorem

Let X be a Banach space and D a closed subset of X. We write $T \in S_\omega(D)$ if T is a strongly continuous semigroup of type ω and, that is $||T(t)||_{Lip} \leq e^{\omega t}$. An operator G is the *infinitesial generator* of T if

$$Gf = \lim_{t \downarrow 0} \frac{T(t)f - f}{t} \tag{3.1}$$

where $\mathcal{D}(G)$ consists of all functions $f \in D$ for which this limit exists. In the nonlinear theory the operator G may be empty. For an example let $V(t)$ be a unitary group on a Hilbert space H, and let A be the generator of the unitary group on H. Suppose that A is unbounded, A^{-1} exists as a bounded operator and choose $f \notin \mathcal{D}(A)$. Set $D = \{U(t)f : t \in \mathbb{R}\}$. Clearly, D is a closed subset of H but not a linear space and $U|_D \in S_0(D)$, but $\mathcal{D}(A) \cap D = \phi$. Thus $G = \phi$.

If C is a subset of X, we define

$$|C| = \inf_{f \in C} ||f||. \tag{3.2}$$

Let $A \subseteq X \times X$ is a multivalued function from X to X, that is $Af = \{g \in X : (f,g) \in A\}$. The domain of A, $\mathcal{D}(A)$, is

$\{f : Af \neq \emptyset\}$ and the range $R(A) = \cup\{Af : f \in \mathcal{D}(A)\}$. If $B \subseteq X \times X$ is a multivalued function and $\lambda \in \mathbb{R}$ or $\mathbb{C}$, then $A + B = \{(f, g + h) : (f, g) \in A \text{ and } (f, h) \in B\}$, $\lambda A = \{(f, \lambda g) : (f, g) \in A\}$ and $A^{-1} = \{(g, f) : (f, g) \in A\}$. We identify functions with their graphs.

An operator A on X is *dissipative* if $||(I - \lambda A)^{-1}||_{Lip} \leq 1$; A is *m-dissipative* if A is dissipative and $R(I - \lambda A) = X$ holds for some $\lambda > 0$. We say A is *essentially m-dissipative* if A is dissipative and $\overline{R(I - \lambda A)} = X$, for some $\lambda > 0$.

Let X^* be the dual space of X. The normalized *duality map* $J : X \to 2^{X^*}$ is defined by

$$J(x) = \{x^* \in X^* : \langle x, x^* \rangle = ||x||, ||x^*|| = 1\}. \tag{3.3}$$

Hence, $\varphi \in J(f)$ iff $\langle f, \varphi \rangle = ||f||\ ||\varphi|| = ||f||$. If $X = H$, a Hilbert space, then $J = \frac{1}{||\cdot||} I$ (except at the origin) where $I : H \to H$ is the identity. If $X = C[0,1]$, the real space, then for $f \in X$, there is a $\xi \in [0,1]$ such that $||\varphi|| = \pm\varphi(\xi)$. Then $\varphi \in J(f)$ if $\langle f, \varphi \rangle = ||f|| = \pm f(\xi)$, hence $\varphi = \pm\delta_\xi$. From the linear theory we have the following lemma.

Lemma 3.1: *Let $f, g, \in X$. Then $||f|| \leq ||f - \alpha g||$ for all $\alpha > 0$ iff there exists $\varphi \in J(f)$ satisfying $Re\langle g, \varphi \rangle \leq 0$.*

If we apply this lemma to the nonlinear case we obtain an alternate definition of dissipativity.

Lemma 3.2: *Let A be a (possibly) multivalued function on $X \times X$ into X. Then A is dissipative iff for all $u_1, u_2 \in \mathcal{D}(A)$ and with $v_i \in Au_i$, there exists $\varphi \in J(u_1 - u_2)$ such that*

$$Re\langle v_1 - v_2, \varphi \rangle \leq 0.$$

This lemma follows easily from Lemma 3.1 by noting that A is dissipative iff

$$||(I - \lambda A)^{-1} h_1 - (I - \lambda A)^{-1} h_2|| \leq ||h_1 - h_2||$$

for all $h_1, h_2 \in R(I - \lambda A)$ and for all $\lambda > 0$. Setting $u_i = (I - \lambda A)^{-1} h_i$, and $v_i \in Au_i$, we see that

$$\begin{aligned} ||u_1 - u_2|| &\leq ||u_1 - \lambda v_1 - u_2 + \lambda v_2|| \\ &= ||\,(u_1 - u_2) - \lambda(v_1 - v_2)\,||, \end{aligned}$$

so Lemma 3.2 follows from Lemma 3.1.

One can also show that if A is m-dissipative, then $R(I - \lambda A) = X$ for all $\lambda > 0$.

Before beginning our outline of the proof of the Crandall-Liggett Theorem we make a few remarks. First, we shall allow for multivalued operators A. In addition the Crandall-Liggett Theorem remains valid if we only assume that $A - \omega I$ is dissipative for some $\omega \in \mathbb{R}$ and that there exists $\lambda_0 > 0$ such that $R(I - \lambda A) \supset \overline{\mathcal{D}(A)}$ for all $0 < \lambda \leq \lambda_0$. In this case A generates $T \in S_\omega(\overline{\mathcal{D}(A)})$ via the formula (1.13).

Outline of the Proof of The Crandall-Liggett Theorem.

We outline the proof in the case $\omega = 0$. We assume that $||(I - \lambda A)^{-1}||_{Lip} \leq 1$, and we must show that

$$J^n_{t/n} f = \left(I - \frac{t}{n} A\right)^{-n} f \to T(t) f$$

as $n \to \infty$. The main step is thus to show that $\left\{J^n_{t/n} f\right\}$ is a Cauchy sequence for $f \in \mathcal{D}(A)$ So for $n \geq m$ and $0 < \mu \leq \lambda$, we define $J_\lambda = (I - \lambda A)^{-1}$ and $a_{m,n} = ||J^n_\mu f - J^m_\lambda f||$.

Recall the resolvent identity for linear operators A: for $\delta, \gamma > 0$,

$$(\gamma I - A)^{-1} - (\delta I - A)^{-1} = (\delta - \gamma)(\gamma I - A)^{-1}(\delta I - A)^{-1}.$$

Its analogue for nonlinear operators is

$$J_\lambda f = J_\mu \left(\frac{\mu}{\lambda} f + \frac{\lambda - \mu}{\lambda} J_\lambda f\right)$$

for $\lambda, \mu > 0$. Then

$$a_{m,n} = ||J^n_\mu f - J^m_\lambda f|| = ||J^n_\mu f - J_\mu \left(\alpha J^{m-1}_\lambda f + \beta J^m_\lambda f\right)||$$

where $\alpha = \frac{\mu}{\lambda}$ and $\beta = \frac{\lambda - \mu}{\lambda}$. Hence

$$a_{m,n} \leq ||(\alpha + \beta) J^{m-1}_\mu f - \left(\alpha J^{m-1}_\lambda f + \beta J^m_\lambda f\right)||$$

since $||J_\mu||_{Lip} \leq 1$. Thus, we have

$$a_{m,n} \leq \alpha a_{m-1,n-1} + \beta a_{m,n-1}. \tag{3.4}$$

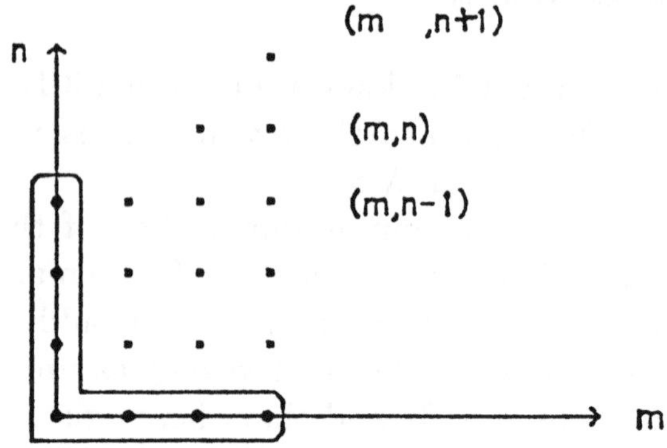

Figure 5.

Identify $a_{m,n}$ with the weight associated with the point (m,n). As the above figure indicates, if we start with $a_{m,n}$ and repeatedly apply (3.4), we will end up with terms involving $a_{k,0}$ or $a_{0,j}$ together with certain coefficients. More specifically

$$a_{m,n} \leq \sum_{j=0}^{m-1} C_{m,n,j}\, a_{j,0} + \sum_{j=0}^{n-m} D_{m,n,j}\, a_{0,j}$$

where $C_{m,n,j}$ and $D_{m,n,j}$ are combinatorial-type coefficients. We estimate the remaining coefficients by

$$\begin{aligned} a_{0,j} = \|J_\mu^j f - f\| &\leq \sum_{k=0}^{j-1} \|J_\mu^{k+1} f - J_\mu^k f\| \\ &\leq j \sum_{k=0}^{j-1} \|J_\mu f - f\| \\ &\leq j\mu \|Af\|. \end{aligned}$$

Here, for simplicity, we have assumed that A is single-valued. The last step follows from

$$\begin{aligned} \|J_\mu f - f\| &= \|J_\mu(u - \mu A u) - J_\mu(f - \mu A f)\| \\ &\leq \|u - \mu A u - f + \mu A f\| \leq \mu \|Af\| \end{aligned}$$

if $(I - \mu A)^{-1} f = u$. We can bound the weights $a_{j,0}$ similarly. □

Again we note that the range condition maybe weakened to $\overline{R(I - \lambda A)} = X$ or $\overline{R(I - \lambda A)} \supseteq \mathcal{D}(A)$, and the Crandall-Liggett theorem still gives a semigroup via the formula

$$T(t) = \lim_{n\to\infty} (I - \frac{t}{n}\bar{A})^{-n}.$$

We also note that the Crandall-Liggett theorem will give local solutions where they exist. For example, consider the equation $u' = u^2$ with the initial condition $u(0) = \frac{1}{c}$ with $c > 0$. Then $u(t) = \frac{-1}{t-c}$, so there is a smooth solution which exists up to time c where c can be arbitrarily small. The Crandall-Liggett theorem gives the formula for this solution, where it exists.

In the linear theory, we can assume

$$||T(t)|| \leq Me^{\omega t}$$

for some $M > 1$ and $\omega > 0$. There is no analogue fo this in the nonlinear theory. Recall that the Crandall-Liggett proof required "stripping off" powers of J_μ^n one at a time. If $||J_\mu^n||_{Lip} \leq M$ with $M > 1$, then these powers of M build up, and the proof breaks down. In fact, the Lipschitz condition for general ω and $f \in \mathcal{D}(A)$ is

$$||T(t+h)f - T(t)f|| \leq e^{\omega t} e^{\omega + h} |Af|.$$

We write $A \in G_\omega$ for $\omega \in \mathbb{R}$ if $A - \omega I$ is closed and dissipative and there is a $\lambda_0 > 0$ such that $R(I - \lambda A) \supset \mathcal{D}(A)$ for $0 < \lambda \leq \lambda_0$.

We write $A \in G_\omega$if $A - \omega I$ is dissipative and $R(I - \lambda A) = X$ for small $\lambda > 0$. Then A generates a semigroup of type W by the Crandall-Liggett theorem, and A is closed. If $\underline{\{T(t)\}}$ is the semigroup generated by $A \in G_\omega^m$, we write $T \in \mathcal{S}_\omega((\mathcal{D}(A))$. Clearly, if B is a globally Lipschitzian operator, then $B \in G^m_{||B||_{Lip}}$, and

$$A + B \in G^m_{\omega + ||B||_{Lip}}.$$

We shall return to perturbation theorems in a later section.

4. The Notion of Solution

In this section we discuss the sense in which the semigroup given by the Crandall-Liggett theorem determines a solution of the abstract Cauchy problem

$$\begin{aligned} u'(t) &\in A(u(t)) \qquad & 0 \leq t < \infty \\ u(0) &= f & f \in \mathcal{D}(A)\,. \end{aligned} \qquad (ACP)$$

A function u is a *strong solution* of (ACP) if u is locally absolutely continuous from $\mathbb{R}^+$ in $X, u(t) \in \mathcal{D}(A)$ for almost every t, $u'(t) \in Au(t)$ for almost every t, and $u(0) = f$.

The following theorem shows that if (ACP) has a strong solution, then it is unique and the semigroup of the Crandall-Liggett theorem gives it.

Theorem 4.1: *Suppose $A \in G_\omega^m$. Consider (ACP) for $f \in \mathcal{D}(A)$. Then for a function $u : \mathbb{R}^+ \to X$ the following are equivalent:*

(i) u is a strong solution of (ACP).

(ii) u is strongly differentiable a.e., and $u(t) = T(t)f$ for all $t \geq 0$.

Corollary: *Suppose $A \in G_\omega^m$ and suppose X is reflexive, then (ACP) has a strong solution which is unique.*

The proof follows from the Fundamental Theorem of Calculus and the preceding theorem. Note that if we assume only $A \in G_\omega$, then by the Crandall-Liggett theorem A generates a semigroup via the exponential formula, so for $f \in \overline{\mathcal{D}(A)}, T(t)f$ is well-defined. Theorem 4.1 remains valid for the solution $u(t)$ of (ACP) for $0 \leq t < T$. We also note that the Corollary remains valid whenever X has the Radon-Nikodym property. However, many spaces which are important in applications do not have the Radon-Nikodym property, such as $L^1(\mathbb{R})$ or $C[0, T]$ equipped with the supremum norm. In nonreflexive spaces it is possible for (ACP) to have a "solution" which is not weakly differentiable at a single point.

We want a notion of solution in a general Banach space X so that we have existence, uniqueness and continuous dependence on initial data for (ACP). There are two such notions of solution; these notions are equivalent when $A \in G_\omega$. The first notion, based on differences schemes, is due to Kobayashi [25] with substantial extensions and improvements due to Evans [14]. The second notion of solution (which is the first historically) is due to Bénilan [5]; it is based on a family of inequalities.

We consider the abstract inhomogeneous Cauchy problem

$$\begin{cases} u'(t) \in Au(t)+g(t) & 0 \leq t \leq T \\ u(0) = f. \end{cases} \qquad (AICP)$$

Here $T > 0$, possibly $T = +\infty$. Let $g \in L^1([0, T]; X)$ and $f \in \overline{\mathcal{D}(A)}$. For each $n \in \mathbb{N}$, we consider a partition $0 = t_0^n < t_1^n < \ldots < t_{N(n)}^n = T$, and let $x_k^n \in \mathcal{D}(A)$, $g_k^n \in X$ for $n \in \mathbb{N}$ and

$k \in \{0, \ldots N(n)\}$. A *backward difference scheme* for (AICP) is a series of triples $\{(t_k^n, x_k^n, g_k^n) : n \in \mathbb{N} \text{ and } k \in \{0, \ldots, N(n)\}\}$ which satisfies the following properties.

$$\frac{x_k^n - x_{k-1}^n}{t_k^n - t_{k-1}^n} \in Ax_k^n + g_k^n \tag{4.1}$$

for $n \in \mathbb{N}$ and $k = 1, \ldots, N(n)$;

$$\lim_{n \to \infty} \max_k \left(t_k^n - t_{k-1}^n\right) = 0; \tag{4.2}$$

$$\lim_{n \to \infty} ||x_0^n - f|| = 0; \tag{4.3}$$

$$\int_0^T ||g_n(t) - g(t)||dt \to 0 \text{ as } n \to \infty. \tag{4.4}$$

where

$$g_n(t) = \begin{cases} g_k^n & \text{for } t \in \left(t_{k-1}^n, t_k^n\right] \\ g_0^n & \text{for } t = 0. \end{cases}$$

Now we define

$$u_n(t) = \begin{cases} x_k^n & t \in \left(t_{k-1}^n, t_k^n\right] \\ x_0^n & t = 0. \end{cases}$$

If $u_n(t) \to u(t)$ uniformly as $n \to \infty$, and $u(t)$ is continuous, then we say $u(t)$ is the *limit solution* of (AICP) on $[0, T]$.

Notice that (4.1) may be rewritten as

$$x_k^n = (I - \delta_k^n A)^{-1} \left(x_{k-1}^n + \delta_k^n g_k^n\right) \tag{4.5}$$

where $\delta_k^n = t_k^n - t_{k-1}^n$ is the length of the kth time step. If $A \in G_\omega^m$, then the range condition $R(I - \lambda A)$ insures that the mapping in (4.5) is well-defined.

For limit solutions existence is easy to prove, but uniqueness is difficult. In addition it is not easy to say in the sense which limit solutions are differentiable. Often we require other techniques to prove additional regularity.

To motivate the second notion of solution, we first consider the case of X a Hilbert space. Let u be a strong solution of

(AICP), and suppose that $A - wI$ is dissipative and single-valued. Then if $x \in \mathcal{D}(A)$

$$\begin{aligned}
\frac{1}{2}\frac{d}{ds}||u(s)-x||^2 &= \frac{1}{2}\frac{d}{ds}\langle u(s)-x, u(s)-x\rangle \\
&= Re\langle u'(s), u(s)-x\rangle \\
&= Re\langle Au(s)+g(s), u(s)-x\rangle \\
&= Re\langle Au(s)-Ax, u(s)-x\rangle \\
&\quad + Re\langle Ax+g(s), u(s)-x\rangle \\
&\leq \omega||u(s)-x||^2 \\
&\quad + Re\langle Ax+g(s), u(s)-x\rangle.
\end{aligned} \tag{4.6}$$

That the second equality in (4.6) holds is an easy lemma. Integrating (4.6) from $s = r \geq 0$ to $s \in [r, T]$ we get

$$\begin{aligned}
\frac{1}{2}||u(t)-x||^2 \leq \frac{1}{2}||u(r)-x||^2 &+ \omega \int_r^t ||u(s)-x||^2 ds \\
&+ \int_r^t Re\langle g(s)+Ax, u(s)-x\rangle ds.
\end{aligned} \tag{4.7}$$

We say u is an *integral solution* of (AICP) if $u(t)$ satisfies (4.7) for $0 \leq r \leq t \leq T$. In order to extend this notion to Banach spaces note that the equality $\frac{1}{2}\frac{d}{ds}||u(s)-x||^2 = Re\langle u'(s), u(s)-x\rangle$ is true in a general Banach space in the form $\frac{1}{2}\frac{d}{ds}||u(s)-x||^2 = Re\langle u'(s), \phi\rangle$ for any $\phi \in J(u(s)-x)$.

Clearly integral solutions are not unique as the next example shows.

Consider (AICP) with the operator $A = \Delta$ and $g \equiv 0$, on the space $X = L^2(\Omega)$ and $\mathcal{D}(A) = C_c^\infty(\Omega)$. Let Ω be a smooth bounded domain $\mathbb{R}^n$ and suppose $A_D = \Delta$ with Dirichlet boundary conditions and $A_N = \Delta$ with Neumann boundary conditions. The solutions, $\exp(tA_D)f$ and $\exp(tA_N)f$, of these problems are different, but both are integral solutions of (ACP). Hence, we must require more if our new notion of solution is to be unique.

For $x, y \in X$, we define

$$\langle x, y\rangle_s = \sup\{Re\langle x, \phi\rangle : \phi \in J(y)\}. \tag{4.8}$$

Suppose $A - \omega I$ is dissipative. Then a continuous function $u : [0,T] \to X$ for X a Banach space is an *integral solution* of (AICP) on $[0,T]$ if $u(0) = f$ and

$$\begin{aligned}\frac{1}{2}||u(t)-x||^2 \le \frac{1}{2}||u(r)-x||^2 + \omega\int_r^t ||u(\tau)-x||^2 d\tau. \\ + \int_r^t \langle g(\tau)+y, u(\tau)-x\rangle_s\end{aligned} \tag{4.9}$$

holds for all $0 \le r \le t \le T$ all $x \in \mathcal{D}(A)$ and for all $y \in Ax$. A continuous function $u : P[0,T] \to X$ is called a *mild solution* of (AICP) on $[0,T]$ if $u(0) = f$ and for all integral solutions of

$$\begin{cases} v' = Av + h(t) & \text{on } [0,T] \\ v(0) = f \end{cases} \tag{4.10}$$

we have

$$\begin{aligned}\frac{1}{2}||u(t)-v(t)||^2 \le \frac{1}{2}||u(r)-v(r)||^2 \\ + \int_r^t \langle g(\tau)-h(\tau), u(\tau)-v(\tau)\rangle_s d\tau\end{aligned} \tag{4.11}$$

for $0 \le r \le t \le T$.

One can easily show that if u and v are strong solutions, then (4.11) holds. Also, if $y \in Ax$ and if we set $h(t) \equiv -y$ and $v(t) \equiv x$, it is clear that a mild solution is an integral solution. Mild solutions satisfy a Gronwall-type inequality; hence, uniqueness, for mild solutions is easy to prove. However, existence is difficult.

Theorem 4.2: *Consider the abstract inhomogeneous Cauchy problem (AICP). Suppose $A - \omega I$ is m-dissipative $g \in L^1([0,T]; \overline{\mathcal{D}(A)})$ and $f \in \overline{\mathcal{D}(A)}$. Then (AICP) has a unique limit solution on $[0,T]$ which coincides with its unique mild solution on $[0,T]$. When $g(t) \equiv 0$, the solution is given by $u(t) = T(t)f$, where $T(t)$ is the semigroup generated by $A-\omega I$ via the Crandall-Liggett theorem. The solution $u(t)$ exists for all $t \in \mathbb{R}^+$ provided $g \in L^1([0,T]; \overline{\mathcal{D}(A)})$ for all $T > 0$.*

We close this section with some brief observations concerning the question of differentiability of the semigroup.

Suppose A generates a semigroup of type ω. The *Yosida approximation* A_λ of A is defined by

$$A_\lambda f = \lambda^{-1}(J_\lambda - I)f \tag{4.12}$$

for $\lambda > 0$ and $f \in \overline{\mathcal{D}(A)}$. We define the functional

$$N(f) = \lim_{\lambda \downarrow 0} ||A_\lambda f|| \tag{4.13}$$

which exists in $[0, \infty]$. It is not difficult to see that $N(f) < \infty$ may be viewed as the condition that the semigroup $T \in \mathcal{S}_\omega(\overline{\mathcal{D}(A)})$ applied to f satisfies a Lipschitz condition with respect to t. More specifically, we can show that

$$N(f) = \lim_{t \downarrow 0} t^{-1} ||T(t)f - f||; \tag{4.14}$$

that is, one side of (4.14) is finite iff the other is and equality holds.

We define the *Favard class* (or the *generalized domain*) of A to be

$$\widehat{\mathcal{D}}(A) = \{f \in \overline{\mathcal{D}(A)} : N(f) < \infty\}.$$

We write $A \in G_\omega^{\max}$ if $A \in G_\omega$ and $A - \omega I$ has no proper dissipative extension in X. Clearly,

$$G_\omega^m \subset G_\omega^{\max} \subset G_\omega,$$

and if $A \in G_\omega$, then A may be extended to $A_0 \in G_\omega^{\max}$ by Zorn's lemma. It is also clear that

$$\mathcal{D}(A) \subseteq \widehat{\mathcal{D}}(A) \subseteq \overline{\mathcal{D}(A)}.$$

The following theorem summarizes the differentiability properties of the semigroup.

Theorem 4.3: *Let $A \in G_\omega^{\max}$, and let $T \in \mathcal{S}_\omega(\overline{\mathcal{D}(A)})$ which A generates.*

(i) For all $t \in \mathbb{R}^+$,

$$T(t)(\widehat{\mathcal{D}}(A)) \subseteq \widehat{\mathcal{D}}(A).$$

(ii) If $t_n \to 0$ as $n \to \infty$ and

$$g = w\text{ - }\lim_{n \to \infty} \frac{T(t_n)f - f}{t_n}$$

where $f \in \mathcal{D}(A)$, *then* $g \in Af$ *and*

$$N(f) = ||g|| = \lim_{n\to\infty} t_n^{-1}||\ T(t_n)f - f||.$$

(iii) If X *is reflexive, then* $\mathcal{D}(A) = \widehat{\mathcal{D}}(A) = \{f \in \overline{\mathcal{D}(A)} :$ *as* $t \to 0, ||T(t)f - f|| = O(t)\}$ *and for all* $f \in \mathcal{D}(A)$,

$$N(f) = \lim_{t\downarrow 0} t^{-1}||T(t)f - f||.$$

In a nonreflexive Banach space, (i) can fail to hold if we replace $\widehat{\mathcal{D}}(A)$ by $\mathcal{D}(A)$.

5. Approximation and Perturbation Theory

The following hypothesis will be assumed throughout this section

Assume

$A_n - \omega I$ *is dissipative and* $R(I - \lambda A_n) \supseteq \overline{\mathcal{D}(A_n)}$ *for each* (HYP)
$n \in \mathbb{N}_0$ *and for* $0 < \lambda < \lambda_0$, *where* λ_0 *is independent of* n.

In particular (HYP) implies that $A_n \in G_\omega$ for each $n \in \mathbb{N}_0$. Let $T_n \in \mathcal{S}_\omega(\overline{\mathcal{D}(A_n)})$ be the semigroup determined by A_n via the exponential formula.

We are interested in theorems which give continuous dependence of the solution $u_n(t)$ of

$$\begin{cases} u_n'(t) \in A_n u_n(t) \\ u_n(0) = f_n \end{cases} \qquad (ACP)_n$$

on both the initial data f_n and the operators A_n.

Theorem 5.1: *Let* (HYP) *hold. Suppose that for each* $g_0 \in \mathcal{D}(A_0)$, *there is a* $g_n \in \mathcal{D}(A_n)$ *such that* $g_n \to g_0$ *as* $n \to \infty$. *Assume that if* $f_n \in \overline{\mathcal{D}(A_n)}$ *and* $f_n \to f_0$, *then*

$$(I - \lambda A_n)^{-1} f_n \to (I - \lambda A_0)^{-1} f_0 \qquad \textit{as } n \to \infty \tag{5.1}$$

for $0 < \lambda < \lambda_1$ *(where* λ_1 *is independent of* n*). Then*

$$\lim_{n\to\infty} T_n(t)f_n = T_0(t)f_0 \tag{5.2}$$

whenever $f_n \in \overline{\mathcal{D}(A_n)}$ *and* $f_n \to f_0$*. Moreover the convergence is uniform for* t *in compact subsets of* $\mathbb{R}^+$*.*

Theorem 5.2: *Let* (HYP) *hold, and suppose that* $\overline{\mathcal{D}(A_0)} \subseteq \overline{\mathcal{D}(A_n)}$ *for all* $n \in \mathbb{N}_0$ *and for all* $f_0 \in \overline{\mathcal{D}(A_0)}$*,*

$$(I - \lambda A_n)^{-1} f_0 \to (I - \lambda A_0)^{-1} f_0 \qquad \text{as } n \to \infty \tag{5.3}$$

for $0 < \lambda < \lambda_1$ *(where* λ_1 *is independent of* n*). Then*

$$\lim_{n\to\infty} T_n(t)f_n = T_0(t)f_0$$

whenever $f_n \in \overline{\mathcal{D}(A_n)}$ *and* $f_n \to f_0$*, the convergence being uniform for* t *in compact subsets of* $\mathbb{R}^+$*.*

In the *linear* case (that is A_n and $T_n(t)$ are linear and $\overline{\mathcal{D}(A_n)} = X$), the statements

$$T_n(t)f_0 \to T_0(t)f_0 \qquad \text{as } n \to \infty \tag{5.4}$$

and

$$(I - \lambda A_n)^{-1} f_0 \to (I - \lambda A_0)^{-1} f_0 \qquad \text{as } n \to \infty \tag{5.5}$$

are equivalent, while (5.5) implies

$$(I - \lambda A_n)^{-1} f_n \to (I - \lambda A_0)^{-1} f_0 \qquad \text{as } n \to \infty. \tag{5.6}$$

It follows by Theorem 5.1 that (5.6) implies (5.4). Thus, (5.4) is equivalent to (5.6), and the hypotheses in Theorem 5.1 are "best possible" in this sense. That (5.4) implies (5.6) can be seen by looking at

$$\lambda(I - \lambda A_n)^{-1} f_n = (\lambda^{-1} I - A_n)^{-1} f_n = \int_0^\infty e^{\frac{-t}{\lambda}} T_n(t) f_n \, dt$$

and using the dominated convergence thoerem.

In fact we will now show that Theorem 5.2 is a special case of Theorem 5.1. We shall require the following fact: If $A - \omega I$ is dissipative , then

$$||(I - \lambda A)^{-1}||_{Lip} \leq (1 - \lambda\omega)^{-1} \tag{5.7}$$

for $0 < \lambda < \omega^{-1}$.

Now assume the hypotheses of Theorem 5.2, and let $f_n \in \mathcal{D}(A_n), f_n \to f_0$ and $0 < \lambda < \lambda_1$. Then

$$\begin{aligned} &||(I - \lambda A_n)^{-1} f_n - (I - \lambda A_0)^{-1} f_0|| \\ &\quad \leq ||(I - \lambda A_n)^{-1} f_n - (I - \lambda A_n)^{-1} f_0|| \\ &\qquad + ||(I - \lambda A_n)^{-1} f_0 - (I - \lambda A_0)^{-1} f_0||. \end{aligned} \tag{5.8}$$

But since $A_n - \omega I$ is dissipative,

$$||(I - \lambda A_n)^{-1} f_n - (I - \lambda A_n)^{-1} f_0|| \leq (1 - \lambda\omega)^{-1} ||f_n - f_0||$$

and by hypothesis (5.3),

$$||(I - \lambda A_n)^{-1} f_0 - (I - \lambda A_0)^{-1} f_0|| \to 0 \text{ as } n \to \infty.$$

Hence, both terms on the right side of (5.8) go to 0 as $n \to \infty$, that is (5.1), and hence the hypothesis of Theorem 5.1, holds.

Next we briefly describe some of the main perturbation theorems for nonlinear semigroups. Our first perturbation theorem is a nonlinear analogue of a standard perturbation theorem in the linear theory.

Theorem 5.3: *Let $A \in G^m_{\omega_1}$, and let $B - \omega_2 I$ be dissipative and single valued with $\mathcal{D}(A) \subset \mathcal{D}(B)$. Suppose there are constants $0 < a < 1$ and $b \geq 0$ such that*

$$||Bf - Bg|| \leq a|Af - Ag| + b||f - g|| \tag{5.9}$$

holds for all $f, g \in \mathcal{D}(A)$. Then

$$A + B \in G^m_{\omega_1 + \omega_2}.$$

Corollary: *If A is m-dissipative and B is globally Lipschitzian, then $A+B \in G^m_{||B||_{Lip}}$.*

Sketch of Proof of Theorem 5.3: Let $0 < a < \frac{1}{2}$. We assume, for simplicity, that A and B are single-valued. Let $\widetilde{A} = A - \omega_1 I$ and $\widetilde{B} = B - \omega_2 I$. Then

$$\begin{aligned} ||\widetilde{B}f - \widetilde{B}g|| &\le ||\; Bf - Bg|| + |\omega_2|\; ||f-g|| \\ &\le a||Af - Ag - \omega_1 f + \omega_1 g|| \\ &\qquad + (b + |\omega_1| + |\omega_2|)||f-g|| \\ &= a||\widetilde{A}f - \widetilde{A}g|| + b_1||f-g||. \end{aligned}$$

Next, $\widetilde{A}+\widetilde{B}$ is dissipative. It remains to show $R(I - \lambda(\widetilde{A}+\widetilde{B})) = X$ for some $\lambda > 0$.

Choose $\lambda > 0$ so large that $(2a + b_1\lambda^{-1}) < 1$ which is possible since $a < \frac{1}{2}$. Then

$$\begin{aligned} &||\widetilde{B}(\lambda I - \widetilde{A})^{-1}f - \widetilde{B}(\lambda - \widetilde{A})^{-1}g|| \qquad (5.10) \\ &\qquad \le a||\widetilde{A}(\lambda I - \widetilde{A})^{-1}f - \widetilde{A}(\lambda I - \widetilde{A})^{-1}g|| \\ &\qquad\qquad + b_1||(\lambda I - \widetilde{A})^{-1}f - (\lambda I - \widetilde{A})^{-1}g|| \\ &\qquad \le a||\lambda(\lambda I - \widetilde{A})^{-1}f - \lambda(\lambda I - \widetilde{A})^{-1}g - (f-g)|| \\ &\qquad\qquad + b_1||(\lambda I - \widetilde{A})^{-1}f - (\lambda I - \widetilde{A})^{-1}g|| \\ &\qquad \le (2a + b_1\lambda^{-1})||f-g||. \end{aligned}$$

Set

$$S_g f = g + \widetilde{B}(I - \lambda\widetilde{A})^{-1}f.$$

Then by (5.10)

$$||S_g f_1 - S_g f_2|| \le (2a + b_1\lambda^{-1})||f_1 - f_2||.$$

Applying the Picard-Banach fixed point theorem we obtain a unique $f_0 \in X$ such that

$$f_0 = g + \widetilde{B}(I - \lambda\widetilde{A})^{-1}f_0. \qquad (5.11)$$

Setting $u = (I - \lambda\widetilde{A})^{-1}f_0$, we see that (5.11) is equivalent to

$$\lambda u - \widetilde{A}u = g + \widetilde{B}u_0,$$

that is, $g = (\lambda I - (\widetilde{A} + \widetilde{B}))u_0$. Hence $g \in R(\lambda I - (\widetilde{A} + \widetilde{B}))$, and since g was arbitrary, we are done.

For the case $\frac{1}{2} < a < 1$, consider $\widetilde{A} + \alpha\widetilde{B}$ which is m-dissipative for $0 \leq \alpha \leq (2a)^{-1}$ by the preceding argument. The result follows by regarding $\beta\widetilde{B}$ as a perturbation of $\widetilde{A} + \alpha\widetilde{B}$ and iterating this procedure.

Theorem 5.4: *Let $A - \omega_1 I$ be a densely defined linear m-dissipative operator on X. Let $B : X \to X$ be everywhere defined and continuous, and suppose $B - \omega_2 I$ is dissipative. Then $A + B \in G^m_{\omega_1+\omega_2}$. Moreover, if $T(t)$ is the semigroup determined by $A + B$ via the Crandall-Liggett theorem, then the strong infinitesimal generator of $T(t)$ is precisely $A + B$.*

The proof of this theorem is technically difficult, but the idea is simple. Let $S = \{S(t) : t \in \mathbb{R}^+\}$ be the linear semigroup generated by A. Then the main idea is to show that the semigroup T given by the variation of parameters formula

$$T(t)f = S(t)f + \int_0^t S(t-s)BT(s)f ds$$

is well defined, holds globally in t and that its generator is $A + B$.

6. Applications of the Theory.

There are many important examples which can be solved via semigroup theory. We shall consider two of them here.

Spatially Degenerate Parabolic Problems

We consider equations of the forms

$$u_t = Au. \tag{6.1}$$

Such equations are often governed by contraction semigroups on L^p for all $p, 1 \leq p \leq \infty$ if the operator A is a uniformly elliptic second order operator. Here we consider such equations where the operator A is allowed to degenerate at the spatial boundary. We shall give the arguments in one space variable for simplicity. We shall then state, but not prove, some known results in higher dimensions.

We consider the following initial - boundary value problem:

$$\begin{cases} u_t = \varphi(x, u_x)u_{xx} + \psi(x, u, u_x) \\ u(x, 0) = f(x) \\ u(0, t) = \alpha u_x(0, t) \\ u(1, t) = -\beta u_x(0, t). \end{cases} \tag{6.2}$$

Here $x \in [0, 1]$, $t \in \mathbb{R}^+$, and $\alpha, \beta \geq 0$.

We make the following assumptions on φ and ψ:

$$\varphi \in C([0, 1] \times \mathbb{R}),\ \varphi(x, \xi) > 0\ for x \in (0, 1), \xi \in \mathbb{R} \tag{6.2}$$

and $0 \leq \varphi_0(x) \leq \varphi(x, \xi)$ *where* $\varphi_0^{-1} \in L^1[0, 1]$.

$\psi \in C([0, 1] \times \mathbb{R}^2)$, *and there exists* $K > 0$ *such that*

$$|\psi(x, \xi, \eta) - \psi(x, \widehat{\xi}, \eta)| \leq K|\xi - \widehat{\xi}|. \tag{6.3}$$

There exist functions $\mathfrak{M}, \mathfrak{L} : \mathbb{R} \to \mathbb{R}$ *which are continuous and nondecreasing, on* $[0, \infty\}$ *and* $\mathfrak{L}(r) \leq L(1 + |r|)$ *for some* $L > 0$ *and*

$$|\psi(x, \xi, \eta)| \leq \mathfrak{M}(|\xi|)(1 + \varphi_0(x))\mathfrak{L}(|\eta|). \tag{6.4}$$

We take $X = C[0, 1]$ equipped with the supremum norm. We consider the abstract Cauchy problem

$$\begin{cases} u_t = Au(t) \\ u(0) = f(x) \end{cases} \tag{6.5}$$

where the operator A is defined by

$$(Au)(x) = \varphi(x, u')u'' + \psi(x, u, u')$$

with domain

$$\mathcal{D}(A) = \{u \in C^2(0, 1) \cap C^1[0, 1] : Au \in C[0, 1] \\ \text{and } u(0) = \alpha u'(0), u(1) = -\beta u'(1)\}.$$

The following claim follows from the second derivative test.

Claim: *If* $u \in C^2(0,1) \cap C^1[0,1]$ *satisfies the boundary conditions in (6.2), then* $\|u\| \neq \pm u(\xi)$ *for* $\xi = 0, 1$*. Hence, if* $\|u\| = \pm u(\xi)$*, then* $0 < \xi < 1$ *and*

$$u'(\xi) = 0, \mp u''(\xi) \geq 0.$$

Our strategy is to show that $A - \omega I$ is m-dissipative for some $\omega > 0$, so that by the Crandall-Liggett theorem and the results of Bénilan-Kobayashi we obtain a unique mild solution of (6.5). Hence, in a certain sense, we have solved (6.1). Additional regularity results from the mild solution follow from results of [13]. We shall state those later.

Lemma 1: *$A - \omega I$ is dissipative for $\omega \geq K$ (where K is as in (6.3)).*

Proof: We wish to show that

$$\langle Au - Av, \phi \rangle \leq \omega \|u - v\|$$

for $\phi \in J(u-v)$. Since we are using the supremum norm, $\phi \in J(w)$ iff $\phi = \pm\delta_\xi$ where $\|w\| = \pm w(\xi)$. Suppose that $\|u - v\| = (u - v)(\xi)$. To avoid trivialities, we assume that $u \neq v$. By the claim we have $0 < \xi < 1, u'(\xi) = v'(\xi)$ and $u''(\xi) \leq v''(\xi)$

$$\begin{aligned}\langle Au - Av, \phi \rangle &= (\varphi(\xi, u'(\xi))u''(\xi) - \varphi(\xi, v'(\xi)))v''(\xi) \\ &\quad + \psi(\xi, u(\xi), u'(\xi)) - \psi(\xi, v(\xi), v'(\xi)) \\ &= \varphi(\xi, u'(\xi))[u''(\xi) - v''(\xi)] \\ &\quad + \psi(\xi, u(\xi), u'(\xi)) - \psi(\xi, v(\xi), u'(\xi)) \\ &\leq K(u - v)(\xi) \leq \omega\|u - v\|\end{aligned}$$

if $\omega \geq K$.

Theorem 6.1: *$A - \omega I$ is m-dissipative for $\omega \geq K$.*

Proof: First we prove the result in the case A is dissipative, that is, $\omega = 0$. This holds, for example, if $\psi(x, \xi, \eta)$ is either independent of ξ or is nondecreasing in ξ.

By Lemma 1, it suffices to show $R(I - \lambda A) = C[0,1]$ for λ sufficiently large, that is given $h \in C[0,1]$ we must find a function $u \in \mathcal{D}(A)$ with $u = (I - \lambda A)^{-1}h$. Hence, u must satisfy

$$\begin{cases} u - \lambda\varphi(\cdot, u')u'' - \lambda\psi(\cdot, u, u') = h \\ u(0) = \alpha u'(0), u(1) = -\beta u'(1). \end{cases} \tag{6.6}$$

Let G be the Green's function for

$$\begin{cases} u'' = h \\ u(0) = \alpha u'(0), u(1) = -\beta u'(1). \end{cases} \tag{6.7}$$

Define the operator $S : C[0,1] \to C[0,1]$ by

$$Su(x) = \int_0^1 G(x,y) \frac{u(y) - h(y) - \lambda\psi(y, u(y), u'(y))}{\lambda\varphi(y, u'(y))} dy. \tag{6.8}$$

Then solving (6.6) is equivalent to finding a fixed point u of S. To accomplish this we use the next theorem.

Theorem 6.2: ***(Schauder Fixed Point Theorem)*** *Let X be a Banach space. If $T : X \to X$ is continuous, compact and leaves some closed convex set invariant, then T has a fixed point in X. That is, there exists $x \in X$ such that $Tx = x$.*

Clearly, $S : C[0,1] \to C^1[0,1]$ and S is continuous. Let $Su = w$. Consider the case of sublinear growth of $\psi(x, \xi, \eta)$ in η, that is, suppose

$$|\psi(x, \xi, \eta)| \leq \mathfrak{M}(|\xi|)(1 + \varphi_0(x))\mathfrak{L}(|\eta|) \tag{6.9}$$

where $\frac{\mathfrak{L}(\eta)}{\eta} \to 0$ as $|\eta| \to \infty$. Then

$$\begin{aligned} ||w|| \leq{} & ||G||\lambda^{-1}||\varphi_0^{-1}||_1(||u|| + ||h||) \\ & + ||G||\mathfrak{M}(||u||)(1 + ||\varphi_0^{-1}||_1)\mathfrak{L}(||u'||) \end{aligned} \tag{6.10}$$

follows from (6.3), (6.8) and (6.9). Also, since

$$w'' = \frac{u - h - \lambda\psi(\cdot, u, u')}{\lambda\varphi(\cdot, u')},$$

we see that

$$\begin{aligned} ||w''||_1 \leq{} & \lambda^{-1}||\varphi_0^{-1}||_1(||u|| + ||h||) \\ & + \mathfrak{M}(||u||)(||\varphi_0^{-1}||_1 + 1)\mathfrak{L}(||u'||). \end{aligned} \tag{6.11}$$

Claim: If $f \in C^1(0,1) \cap C^2[0,1]$ with $f'' \in L^1[0,1]$ then

$$||f'|| \leq 4||f|| + ||f''||_1. \tag{6.12}$$

The proof of this claim is easy; it can be found in [13].

Moreover it follows that $\{(Su)' : u \in B\}$ is a equicontinuous collection if B is bounded in $C^1[0,1]$; hence, by the Arzela-Ascoli theorem, $S : C^1[0,1] \to C^1[0,1]$ is compact.

Choose N_0 so large that $N_0 \geq 5||h||$ and

$$16\mathfrak{M}(||h||)(||\varphi_0^{-1}||_1 + 1)\mathfrak{L}(N_0) \leq N_0.$$

Choose $\lambda > 16(1+||G||)||\varphi_0^{-1}||_1$ and $M_0 = ||h||+N_0$. We consider the closed, convex bounded set

$$B = \{u \in C^1[0,1] : ||u|| \leq M_0, ||u'|| \leq N_0\}.$$

Then by (6.10) and (6.11) we have

$$||w|| \leq \frac{1}{16}(M_0 + ||h||) + \frac{1}{16}N_0 \leq \frac{1}{8}M_0,$$

$$||w''||_1 \leq \frac{1}{16}(M_0 + ||h||) + \frac{1}{16}N_0 \leq \frac{1}{8}M_0.$$

Using (6.12) we see that

$$||w'|| \leq 4||w|| + ||w''||_1 \leq \frac{1}{2}M_0 + \frac{1}{8}M_0 < N_0,$$

whence, by the Schauder fixed point theorem $S(B) \subset B$.

In order to handle the case $\omega > 0$, we must proceed in two steps. First, for each $v \in C[0,1]$, we consider the operator

$$A^v u = \varphi(\cdot, u')u'' + \psi(\cdot, v, u').$$

Clearly, A^v is dissipative; hence, $J_\lambda^v = (I - \lambda A^v)^{-1}$ is well-defined, exists and is single-valued by the preceding argument. Next, for each $h \in C[0,1]$ we consider the operator $S : C[0,1] \to C[0,1]$ defined by

$$S^h v = J_\lambda^v h.$$

Finding a fixed point u of S^h is equivalent to showing that $u - \lambda A u = h$. The details of finding a fixed point of S^h for $0 < \lambda \leq \lambda_0$, where $0 < \lambda_0 < \omega^{-1}$ can be found in [13].

This concludes the proof of Theorem 6.1.

The proof in the case of linear growth in (6.4) is more difficult. For details see [17]. The above results can also be extended to include inhomogeneous periodic and nonlinear boundary conditions. The quasilinear version of problem (6.2) has been studied in [12], [13], [17], [30] for various types of boundary conditions. In these papers questions of regularity for both the semilinear and the addressed quasilinear problem. However, in the quasilinear case only local existence is known.

The case of more rapid degeneracy, that is $\varphi_0(x) \to 0$ arbitrarily fast as x approaches the boundary has been studied by [21] and [31].

Porous Medium Equation

The equation

$$u_t = \Delta(u^{\alpha+1}), \qquad \text{on } \Omega \subset \mathbb{R}^n, \tag{6.13}$$

or more generally the equation

$$u_t = \Delta(\varphi(u)) \qquad \text{on } \Omega \subset \mathbb{R}^n \tag{6.14}$$

governs the flow of a fluid throught a porous medium and can be derived under certain approximations from the Navier-Stokes equations. Here $\alpha > 0, u(t,x) = c\rho(t,x)$ where $c > 0$ and $\rho(t,x)$ represents the density of the fluid at the point $x \in \Omega$ and at time $t \in \mathbb{R}^+$, the pressure is given by $p(t,x) = p_0[\rho(t,x)]^\alpha$ for some $p_0 > 0$. In (6.14) we asume that the mapping $\varphi : \mathbb{R} \to \mathbb{R}$ is either a continuous nondecreasing function or a maximal monotone graph satisfying $0 \in \varphi(0)$. Equation (6.13) may be viewed as a special case of (6.12) where

$$\varphi(s) = \begin{cases} s^{\alpha+1} & \text{for } s \geq 0 \\ \\ -|s|^{\alpha+1} & \text{for } s < 0. \end{cases}$$

We consider the case $\Omega = \mathbb{R}^n$, and the operator A acting on $X = L^p(\mathbb{R}^n)$ where A is the nonlinear operator

$$A = \Delta \circ \varphi$$

with domain

$$\begin{aligned} \mathcal{D}(A) = \{u \in X : \text{there exists } w \in {L^1}_{loc}(\mathbb{R}^n) \\ \text{such that } w(x) \in \varphi(u(x)) \text{ and} \\ \Delta w \in X\}. \end{aligned}$$

Recall that in $L^1(\mathbb{R}^n), \phi \in J(u)$ iff $||\phi|| = 1$ and $< u, \phi >= ||u||_1$; in particular $\phi(u) \in \text{sgn } u$ where

$$\text{sgn } u = \{w \in L^\infty(\mathbb{R}^n) : w(x) \in \text{ sgn } (u(x)) \text{ a.e. } \},$$

and

$$\text{sgn } s = \begin{cases} \{1\} & \text{if } s > 0 \\ [-1,1] & \text{if } s = 0 \\ \{-1\} & \text{if } s < 0. \end{cases}$$

In addition, we note that

$$Au = \{\Delta w : w(x) \in \varphi(u(x)) \text{ for some } u \in \mathcal{D}(A)\};$$

hence, A is single-valued if φ is continuous.

The next theorem is the main result of this section. We break its proof into several parts.

Theorem 6.2: *The operator A is m-dissipative on $L^p(\mathbb{R}^n)$ iff $p = 1$.*

A similar result is valid in Hilbert space context related to $H^{-1}(\Omega)$.

Lemma 2: *A is dissipative on $L^1(\mathbb{R}^n)$.*

In order to prove Lemma 2, we need the following claim.

Claim: *Let $u_i \in \mathcal{D}(A)$ and $Au_i = \Delta w_i$ for $i = 1, 2$. There exists $\phi \in J(u - u_2) \cap J(w_1 - w_2)$.*

Proof of Claim: Without loss of generality we assume that $||u_1 - u_2||_1 = 1$. Since φ is a nondecreasing function or a maximal monotone graph containing the origin and since $w_i(x) \in \varphi(u_i(x))$, we see that $u_1(x) > u_2(x)$ implies $w_1(x) \geq w_2(x)$ and $u_1(x) < u_2(x)$ implies $w_1(x) \leq w_2(x)$. Whence, if we define $\phi((u_1 - u_2)(x)) = \text{sgn } ((u_1 - u_2)(x))$, then $\phi = \pm 1 \in \text{sgn } ((w_1 - w_2)(x))$ as long as $w_1(x) \neq w_2(x)$. If x is such that $w_1(x) = w_2(x)$, let

$$\phi((u_1 - u_2)(x)) = \text{sgn }_0((u_1 - u_2)(x))$$

$$= \begin{cases} 1 & \text{if } u_1(x) > u_2(x) \\ 0 & \text{if } u_1(x) = u_2(x) \\ -1 & \text{if } u_1(x) < u_2(x). \end{cases}$$

Then this ϕ satisfies $\phi \in J(w_1 - w_2) \cap J(u_1 - u_2)$.

Proof of Lemma 2: Let $u_i \in \mathcal{D}(A)$ and $\Delta w_i = Au_i$ for $i = 1, 2$. By the claim we can choose $\phi \in J(u_1 - u_2) \cap J(w_1 - w_2)$. Then $\langle Au_1 - Au_2, \phi \rangle = \langle \Delta w_1 - \Delta w_2, \phi \rangle \leq 0$ since $\phi \in J(w_1 - w_2)$ and Δ is dissipative on L^1(cf. [18]).

In order to complete the proof of Theorem 6.2, it remains to show the range condition; that is we must solve $u - \lambda Au \ni h$ for any $h \in L^1(\mathbb{R}^n)$. We shall actually only prove that A is essentially m-dissipative here. In particular we shall solve $u - \lambda Au \ni h$ for $h \in C_0^\infty(\mathbb{R}^n)$ and for some $\lambda > 0$ sufficiently large (and hence for all $\lambda > 0$).

Set

$$v = \varphi(u) \text{ or } v \in \varphi(u)$$

for $v \in L^1_{loc}(\mathbb{R}^n)$. Define $\beta : \mathbb{R} \to \mathbb{R}$ by $\beta(x) = \frac{1}{\lambda}\varphi^{-1}(x)$. Clearly, φ is a maximal monotone graph satisfying $0 \in \varphi(0)$ iff $\varphi^{-1} = \lambda\beta$ is maximal monotone and satisfies $0 \in \beta(0)$. In addition we see that solving $u - \lambda Au \ni h$ is equivalent to solving

$$-\Delta v + \beta(v) \ni g \tag{6.15}$$

for $g \in C_0^\infty(\mathbb{R}^n)$. We briefly describe the method used to solve (6.15). Let β_m be the Yosida approximation of β more precisely

$$\beta_m = m(I - (I + \frac{1}{m}\beta)^{-1}).$$

It follows that β_m is single-valued, globally Lipschitzian and $\beta_m(0) = 0$. Also,

$$(I - \lambda\beta_m)^{-1} \to (I - \lambda\beta)^{-1} \quad \text{as } m \to \infty \tag{6.16}$$

for all $\lambda > 0$. Hence, by the Corollary to Theorem 5.3., $\Delta - \beta_m$ is m-dissipative on $L^1(\mathbb{R}^n)$.

So we have shown that for all $\varepsilon > 0$ there exists $u_{m,\varepsilon}$ which satisfies

$$\varepsilon u_{m,\varepsilon} - \Delta u_{m,\varepsilon} + \beta_m(u_{m,\varepsilon}) = g. \tag{6.17}$$

The main strategy is to let $m \to \infty$ and $\varepsilon \to 0$, but the order in which we do this is very important. We shall show there exists a subsequence of $\{u_{m,\varepsilon}\}$ which converges to limit v which satisfies (6.15).

Fix $\varepsilon > 0$. Multiplying (6.17) by $\text{sgn}_0 u_{m,\varepsilon}$ and integrating over $\mathbb{R}^n$, we see that

$$\varepsilon\|u_{m,\varepsilon}\|_1 + \int \beta_m(u_{m,\varepsilon})\, \text{sgn}_0(u_{m,\varepsilon}) \\ \leq \langle g,\ \text{sgn}_0 u_{m,\varepsilon}\rangle \leq \|g\|_1$$

where we have used the dissipativity and the assumptions on β_m. Again, using the equation, we get

$$\|\Delta u_{m,\varepsilon}\|_1 \leq 2\|g\|_1. \tag{6.18}$$

Next, multiplying (6.17) by $|\beta_m(u_{m,\varepsilon})|^{p-1}\ \text{sgn}\,_0\big(\beta_m(u_{m,\varepsilon})\big)$ for $1 \leq p < \infty$, integrating over $\mathbb{R}^n$, and using Hölder's inequality and the monotonicity of β_m, we obtain

$$\|\beta_m(u_{m,\varepsilon})\|_p \leq \|g\|_p. \tag{6.19}$$

Since $\Delta - \beta_m$ is m-dissipative on $L^p(\mathbb{R}^n)$ for $1 \leq p < \infty$, $\varepsilon\|u_{m,\varepsilon}\|_p \leq \|g\|_p$; whence

$$\|\Delta u_{m,\varepsilon}\|_p \leq 3\|g\|_p. \tag{6.20}$$

It follows that $\{u_{m,\varepsilon}\}$ is a bounded sequence in $W^{2,p}(\mathbb{R}^n)$ for $1 < p < \infty$. By the local compactness of the imbedding of $W^{2,p}(\mathbb{R}^n)$ into $L^r(\mathbb{R}^n)$ if r is sufficiently large, we obtain a subsequence, which we again denote by $\{u_{m,\varepsilon}\}$ such that $u_{m,\varepsilon} \to v_\varepsilon \in W^{2,p}_{loc}(\mathbb{R}^n)$, $v_\varepsilon \in W^{2,p}(\mathbb{R}^n)$ and v_ε satisfies

$$\varepsilon v_\varepsilon - \Delta v_\varepsilon + \beta(v_\varepsilon) \ni g. \tag{6.21}$$

Letting $\varepsilon \to 0$ is more difficult. First we note that if v satisfies $-\Delta v + \beta(v) \ni h$, then clearly $u \in \beta(v)$ should be in $L^1(\mathbb{R}^n)$, but in which space should v exist? Take for example the case $\beta \equiv 0$. From Newtonian potential theory we know that if v satisfies $-\Delta v = h$ for $h \in L^1(\mathbb{R}^n)$ then

$$v(x) = c_n \int_{\mathbb{R}^n} \frac{h(y)}{|x-y|^{n-2}}\,dy, \tag{6.22}$$

that is $v = E_n * h$ where $E_n = c_n|\cdot|^{2-n}$. Here $n \geq 3$, which we assume. We would like to use a convolution inequality like Young's inequality to say that $||v||_p \leq ||E_n||_p||h||_1$, but $E_n \notin L^p(\mathbb{R}^n)$ for any $p \in [1,\infty]$. However, E_n is in the *Marcinkiewicz (or weak L^p) space* $L^p_{wk}(\mathbb{R}^n)$, with $p = \frac{n}{n-2}$ which consists of all measurable functions u for which $|||u|||_p < \infty$. Here

$$|||u|||_p = \inf\{K > 0 : \int_\Omega |u(x)|dx \leq K|\Omega|^{\frac{1}{q}} \text{ where } \frac{1}{p} + \frac{1}{q} = 1$$
$$\text{and for all measurable Borel sets } \Omega \text{of finite measure}\}.$$

$L^p_{wk}(\mathbb{R}^n)$ equipped with the norm $|||\cdot|||_p$ is a Banach space. One can show that if $w_1 \in L^p_{wk}(\mathbb{R}^n)$ and $w_2 \in L^1(\mathbb{R}^n)$, then

$$|||w_1 * w_2|||_p \leq |||w_1|||_p||w_2||_1. \tag{6.23}$$

Since $E_n \in L_{wk}^{\frac{n}{n-2}}(\mathbb{R}^n)$, it follows the $v \in L_{wk}^{\frac{n}{n-2}}(\mathbb{R}^n)$ if v solves $-\Delta v = h$. In addition one can show that if $u \in L^p_{wk}(\mathbb{R})$ for some $1 \leq p < \infty$ and $\Delta u \in L^1(\mathbb{R}^n)$, then

$$u = c_n \frac{1}{|\cdot|^{n-2}} * (-\Delta u),$$

$u \in L_{wk}^{\frac{n}{n-2}}(\mathbb{R}^n)$, $|\nabla u| \in L_{wk}^{\frac{n}{n-1}}(\mathbb{R}^n)$ and

$$\begin{cases} |||u|||_{\frac{n}{n-2}} \leq A_n||\Delta u||_1 \\ |||\Delta u|||_{\frac{n}{n-1}} \leq B_n||\Delta u||_1 \end{cases} \tag{6.24}$$

for some constants A_n, B_n independent of u.

In order to let $\varepsilon \to 0$ in (6.21), one must prove that $\{\beta^\varepsilon(u_\varepsilon)\}$ is a precompact sequence in $L^1_{loc}(\mathbb{R}^n)$, where

$$\beta^\varepsilon = \beta + \varepsilon I,$$

and that $||\beta^\varepsilon(u_\varepsilon)||_1 \leq ||g||_1$. Then $\beta^\varepsilon(u_\varepsilon) \to w \in L^1{}_{loc}(\mathbb{R}^n)$ and a.e., and by Fatou's Lemma

$$\int |w| \leq ||g||_1,$$

so $w \in L^1(\mathbb{R}^n)$. Combining (6.20) and (6.24), we see that

$$|||u_\varepsilon|||_{\frac{n}{n-2}} \leq A_n||\Delta u_\varepsilon||_1 \leq \widetilde{A}_n||g||_1$$
$$|||\nabla u_\varepsilon|||_{\frac{n}{n-1}} \leq B_n||\Delta u_\varepsilon||_1 \leq \widetilde{B}_n||g||_1$$

and that $\{u_\varepsilon\}, \{\nabla u_\varepsilon\}$ are precompact in $L^1_{loc}(\mathbb{R}^n)$. Hence, using Fatou's lemma, we obtain a subsequence $\{u_{\varepsilon n}\}$ such that $u_{\varepsilon n} \to v$ in $L^1{}_{loc}(\mathbb{R}^n)$ and a.e., $v \in L_{wk}^{\frac{n}{n-2}}(\mathbb{R}^n)$, $v \in W^{1,1}_{loc}(\mathbb{R}^n), \nabla v \in L_{wk}^{\frac{n}{n-1}}(\mathbb{R}^n), w = \beta(v)$ and v solves

$$-\Delta v + \beta(v) \ni g. \tag{6.25}$$

It follows from a form of the weak maximum principle that v is the unique solution of (6.25). This concludes the proof of Theorem 6.2.

We conclude this section with a brief discussion of some of the qualitative properties of the solution of (6.14).

We have shown that our operator A is m-dissipative on $L^1(\mathbb{R}^n)$, so A generates a semigroup $T(t)$ via the Crandall-Liggett theorem. The semigroup $T(t)$ satisfies the *conservation of mass property* if

$$\int_{\mathbb{R}^n} (T(t)f)(x)dx = \int_{\mathbb{R}^n} f(x)dx \tag{6.26}$$

for all $f \in L^1(\mathbb{R}^n)$. Formally, the porous medium semigroup satisfies (6.26) since

$$\begin{aligned}\frac{d}{dt}\int_{\mathbb{R}^n} (T(t)f)(x)dx &= \int_{\mathbb{R}^n} u_t(t,x)dx \\ &= \int_{\mathbb{R}^n} \Delta\big(\varphi(u(t,x))\big)dx = 0\end{aligned}$$

by the divergence theorem. The following theorem of Crandall and Tartar [10] shows that the conservation property implies the positivity of the semigroup for contraction semigroups on L^1 spaces.

Proposition 6.3: *Let $X = L^1(\Omega, \mu)$ where (Ω, μ) is a measure space; let $C \subseteq X$ such that if $f, g \in C$, then $f \wedge g \in C$. Suppose $S : C \to X$ satisfies*

$$\int_\Omega Sf d\mu = \int_\Omega f d\mu \tag{6.27}$$

for all $f \in C$. Then the following are equivalent:

(i) If $g \in C$, and $f \leq g$ a.e., then $Sf \leq Sg$ a.e.

(ii) $\|Sf - Sg\|_1 \leq \|f - g\|_1$ for all $f, g \in C$.

We shall give the proof that (ii) implies (i), since this is the part we will use. Consider $f, g \in C$ such that $f \leq g$ a.e. Then

$$\begin{aligned} 2\int_\Omega (Sf - Sg)_+ d\mu &= \int_\Omega |Sf - Sg| d\mu + \int_\Omega (Sf - Sg) d\mu \\ &\leq \int_\Omega |f - g| d\mu + \int_\Omega (f - g) d\mu \end{aligned}$$

by (ii) and (6.27). Hence,

$$2\int_\Omega (Sf - Sg)_+ d\mu \leq 2\int_\Omega (f - g)_+ d\mu = 0$$

that is $Sf \leq Sg$ a.e.

Again let us consider the porous medium semigroup $T(t)$ on $L^1(\mathbb{R}^n)$. By Proposition 6.3, for all $t \geq 0$ we have

$$\begin{gathered} f_1 \leq f_2 \text{ a.e. implies} \\ u_1(t,x) = T(t)f_1 \leq T(t)f_2 = u_2(t,x) \text{ a.e. and for all } t > 0 \end{gathered} \tag{6.28}$$

where $u_i(0,x) = f_i(x)$ for $i = 1, 2$.

Now let us specialize to the case

$$\frac{\partial u}{\partial t} = \Delta(u^{\alpha+1}) \qquad t > 0, \alpha > 1, x \in \mathbb{R}^n. \tag{6.29}$$

Here $n \in \mathbb{N}$; also we consider $u^{\alpha+1} = |u|^\alpha u$, the odd extension of $u^{\alpha+1}$ or consider only nonnegative solutions of (6.29). The *Barenblatt solution* of (6.29) is

$$U(t,x;M) = t^{-\gamma}\{a - b|x|^2 t^{-2\delta}\}_+^{1/\alpha}$$

where $M > 0$ is given, $\delta = \frac{1}{2+\alpha n,}\gamma = \frac{1-2\delta}{\alpha}$ $b = \frac{\alpha\delta}{2(\alpha+1)}$ and a is uniquely determined by

$$\int_{\mathbb{R}^n} U(t,x;M) = M \text{ for all } t > 0.$$

Note that if $|x| \geq \left(\frac{a}{b}\right)^{1/2} t^{\delta}$, then $U(t, x; M) = 0$, that is, $U(t, x; M)$ has compact support.

Consider a function u which satisfies (6.29) with

$$u(0, x) = u_0(x) \in L^1(\mathbb{R}^n)$$

where

$$0 \leq u_0(x) \leq U(t_0, x; M_0)$$

for some t_0, M_0 and for all x. Then using (6.28), we see that

$$0 \leq u(t, x) \leq U(t_0 + t, x; M_0)$$

for all $t > 0$ and $x \in \mathbb{R}^n$, that is,

$$\operatorname{supp}(u(t, \cdot)) \subseteq B_R$$

where $R = \left(\frac{a}{b}\right)^{1/2} (t + t_0)^{\delta}$. In particular the solution $u(t, x)$ doesn't become everywhere positive immediately because the support is contained in some ball. Hence, we have shown that the solution of the porous medium equation has a finite speed of propagation. Contrast this result with the limiting case $\alpha = 0$, the linear heat equation. We have

$$u(t, x) = (4\pi t)^{-n/2} \int_{\mathbb{R}^n} e^{\frac{-|x-y|^2}{4t}} u_0(y) dy$$

where $u(0, x) = u_0(x)$. (If $\alpha = 0$, the Barenblatt solution fails.) If $0 \leq u_0(x)$ (and $u_0 \neq 0$ a.e.), then $u(t, x) > 0$ for all $t > 0$, and $x \in \mathbb{R}^n$; that is, the linear heat equation has an infinite speed of propagation.

References

1. D. G. Aronson, The porous medium equation, *Nonlinear Diffusion Problems,* Lecture Notes in Math. 1229, Springer Verlag, Berlin, (1986), 1-46.

2. D. G. Aronson and Ph. Bénilan, Regularite des solutions de l'equations des milieux poreaux dans R^n, *C.R. Acad. Sc. Paris,* 288 (1979), 103-105.

3. D. G. Aronson, L.A. Caffarelli, and J. L. Vazquez, Interface with a corner point in one-dimensional porous medium flow, *Comm. Pure and Appl. Math.,* 38 (1985), 375-404.

4. D. G. Aronson, M. G. Crandall, and L. A. Peltier, Stabilizations of solutions of a degenerate nonlinear diffusion problem, *Nonlinear Anal TMA,* 6 (1982), 1001-1033.

5. Ph. Bénilan, *Equations d'Evolution dans un Espace de Banach Quelconque et Applications,* Ph.D. Thesis, Univ. of Paris (1972).

6. Ph. Bénilan, H. Brezis, and M.G. Crandall, A semilinear elliptic equation in $L^1(R^N)$, *Ann. Scuola Norm. Sup. Pisa,* 2 (1975),523-555.

7. Ph. Bénilan, and M. G. Crandall, The continuous dependence on φ of solutions of $u_t - \Delta\varphi(u) = 0$, *Indiana Univ. Math. J.,* 30 (1981), 161-177.

8. M. G. Crandall and T. M. Liggett, Generation of semigroups of nonlinear transformations on general Banach spaces, *Amer. J. Math.,* 93 (1971), 265-298.

9. M. G. Crandall and A. Pazy, Nonlinear evolution equations in Banach spaces, *Israel J. Math,* 11 (1972), 67-94.

10. M. G. Crandall and L. Tartar, Some relations between nonexpansive and order preserving mappings, *Proc. Amer. Math. Soc.* 78 (1980), 385-390.

11. J. R. Dorroh and G. R. Goldstein, Existence and regularity of solutions of singular parabolic problems, in preparation.

12. J. R. Dorroh and G. R. Goldstein, A singular quasilinear parabolic problem in n dimensions, in preparation.

13. J. R. Dorroh and G. R. Rieder, A singular quasilinear parabolic problem in one space dimension, *J. Diff. Equations,* 91 (1991), 1-23.

14. L. C. Evans, "Nonlinear Evolution Equations in an Arbitrary Banach Space," Math. Res. Center Tech Summary Report No. 1568, Madison (August 1975).

15. L. C. Evans, Nonlinear evolution equations in Banach spaces, *Israel J. Math.,* 26 (1977), 1-42.

16 D. Gilbarg and N. S. Trudinger, *Elliptic Partial Differential Equations of Second Order*, second ed., Springer Verlag, Berlin, 1983.

17. G. R. Goldstein, Nonlinear singular diffusion with nonlinear boundary conditions, *Math. Methods in the Appl. Sciences,* 15 (1993), 1-20.

18. J. A. Goldstein,*Semigroups of Linear Operators and Applications,* Oxford University Press, New York and Oxford, 1985.

19. J. A. Goldstein, *Semigroups of Nonlinear Operators and Applications,* in preparation.

20. J. A. Goldstein and C. Y. Lin, Singular nonlinear parabolic boundary value problems in one space dimension, *J. Diff. Equations* 68 (1987), 429–443.

21. J. A. Goldstein and C.-Y. Lin, Highly degenerate parabolic boundary value problems, *Diff. Int. Eqns.* 2 (1989), 216-227.

22. J. A. Goldstein and C.-Y. Lin, An L^p-semigroup approach to degenerate parabolic bundary value problems, *Ann. Mat. Pura Appl.,* 159 (1991), 211-227.

23. E. Hille, *Functional Analysis and Semi-groups*, Amer. Math. Soc. Coll. Publ. Vol. 31, New York, 1948.

24. E. Hille and R. Phillips, *Functional Analysis and Semi-groups,* Amer. Math. Soc. Coll. Publ. Vol. 31, Providence, R. I., 1957.

25. Y. Kobayashi, Difference approximations of Cauchy problems for quasi-dissipative operators and generation of nonlinear semigroups, *J. Math. Soc. Japan* 27 (1975), 640-655.

26. S. N. Kruzkov, First order quasilinear equations in several independent variables, *Math USSR Sbornik* 10 (1970), 217-243.

27. C.-Y. Lin, Degenerate nonlinear parabolic boundary value problems, *Nonlinear Anal TMA* 13 (1989) , 1303-1315.

28. R. Nagel, *One-Parameter Semigroups of Positive Operators,* Lecture Notes in Math. 1184, Springer-Verlag, Berlin, 1986.

29. A Pazy, *Semigroups of Linear Operators and Applications to partial Differential Equations*,Springer, New York, 1983.

30. G. R. Rieder, Spatially degenerate diffusion with periodic-like boundary conditions, in *Differential Equations with Applications in Biology, Physics and Engineering* (ed. by J. Goldstein, F. Kappel and W. Schappacher), Lecture Notes in Pure and Applied Mathematics, Marcel Dekker, New York (1991), 301-312.

31. A. D. Wentzel, On boundary conditions for multidemensional diffusion processes, *Theory Prob. Appl.*, 4 (1959), 164-177.

32. K. Yosida, On the differentiability and the representation of one-parameter semi-groups of linear operators, *J. Math. Soc. Japan* 1 (1948), 15-21.

A bifurcation problem for point interactions in $L^2(\mathbb{R}^3)$

Wim Caspers and Philippe Clément

1 Introduction

Consider the equation

$$-\Delta u + |u|^{s-1}u = \lambda u \tag{1.1}$$

in $\mathbb{R}^3$ with $1 < s < \frac{3}{2}$ and $u \in L^2(\mathbb{R}^3)$. It is known [8] that this problem has no positive solution for $\lambda \in \mathbb{R}$. In this paper we investigate a slight modification of this nonlinear eigenvalue problem, where we replace the operator $-\Delta$ with domain $H^{2,2}(\mathbb{R}^3)$, by a so-called point interaction centered at zero. Such point interactions (see [1]) can be described as selfadjoint extensions in $L^2(\mathbb{R}^3)$ of the operator $-\Delta$ restricted to $C_0^\infty(\mathbb{R}^3\backslash\{0\})$, smooth functions with compact support in $\mathbb{R}^3\backslash\{0\}$, or equivalently the operator $-\Delta$ restricted to $H_0^{2,2}(\mathbb{R}^3\backslash\{0\})$, that is the space of $H^{2,2}$-functions that are zero in the origin. All selfadjoint extensions are given by a family $\{-\Delta_\tau\}_{\tau\in[0,2\pi)}$, see (2.4) below. These selfadjoint extensions can also be described by means of Fermi pseudopotentials [1], [5].

One interesting property of point interactions is the existence of a positive eigenfunction (also called ground state), for certain values of τ, corresponding to a negative eigenvalue λ_0 of $-\Delta_\tau$. This positive eigenfunction is singular at the origin and behaves like $\frac{C}{r}$ there. It is natural to consider the associated problem

$$-\Delta_\tau u + |u|^{s-1}u = \lambda u \tag{1.2}$$

G. R. Goldstein and J. A. Goldstein (eds.),
Semigroups of Linear and Nonlinear Operations and Applications, 99–108.

for these values of τ. Theorem 5 below, states the existence of nontrivial positive solutions of (1.1). These solutions are continuous on $\mathbb{R}^3$, except at the origin, where they have a singularity of order $\frac{1}{r}$. Multiplying (1.2) by a test function $v \in C_0^\infty(\mathbb{R}^3\backslash\{0\})$ and using the selfadjointness of $-\Delta_\tau$ we get

$$\int_{\mathbb{R}^3} -u\Delta v dx + \int_{\mathbb{R}^3} |u|^{s-1} uv dx = \int_{\mathbb{R}^3} \lambda uv dx. \tag{1.3}$$

So in this way we obtain singular solutions of the elliptic semilinear problem (1.1) in $\mathbb{R}^3\backslash\{0\}$. By considering point interactions in $L^p(\mathbb{R}^3)$ for $\frac{3}{2} < p < 3$ (see remark following Theorem 5) one can even allow $1 < s < 2$. This approach to singular solutions seems to be new. In Theorem 5 we use a bifurcation and a continuation argument to prove the existence of a curve of positive and negative solutions, parametrized by λ between the eigenvalue λ_0 (depending on τ) and 0.

Note that these solutions are radially symmetric. The rescaled function $w(y) = (-\lambda)^{\frac{1}{1-s}} u(\frac{1}{\sqrt{-\lambda}} y)$, $(\lambda < 0)$ satisfies the equation

$$-w''(r) - \frac{2}{r} w'(r) + |w(r)|^{s-1} w(r) = w(r), \quad r > 0.$$

This paper also describes the situation where point interactions associated with $-\Delta$ are replaced by point interactions associated with $-\Delta + V$, with a potential V belonging to some class $\mathcal{C}$ defined below. Full details and complete proofs will appear elsewhere.

2 Point interactions

Usually, see e.g. [1], point interactions (one center, centered at zero) are defined in $L^2(\mathbb{R}^3)$ as selfadjoint extensions of the operator

$$-\Delta : C_0^\infty(\mathbb{R}^3\backslash\{0\}) \subseteq L^2(\mathbb{R}^3) \longrightarrow L^2(\mathbb{R}^3)$$

where $C_0^\infty(\mathbb{R}^3\backslash\{0\})$ denotes the space of infinitely many times differentiable functions with compact support which does not contain the origin. It is well known that all selfadjoint extensions of this restricted Laplacian operator are given by the family of operators $\{-\Delta_\tau\}_{\tau\in(-\pi,\pi]}$ with

$$\begin{aligned} D(-\Delta_\tau) &= \{u_0 + c(\varphi_+ + e^{i\tau}\varphi_-) \ : \ u_0 \in H^{2,2},\ u_0(0) = 0,\ c \in \mathbb{C}\} \\ -\Delta_\tau(u_0 + c(\varphi_+ + e^{i\tau}\varphi_-)) &= -\Delta u_0 + c(i\varphi_+ - ie^{i\tau}\varphi_-), \end{aligned} \tag{2.4}$$

where $\varphi_\pm = \frac{e^{(-\frac{1}{2}\sqrt{2}\pm\frac{1}{2}\sqrt{2}i)|x|}}{4\pi|x|}$ and satisfy $(\mp i - \Delta)_{-2,2}\varphi_\pm = \delta_0$ in the sense of distributions. Here δ_0 denotes the Dirac measure at zero. (Observe that $-\Delta_\pi$ is the usual Laplacian with $D(-\Delta_\pi) = H^{2,2}(\mathbb{R}^3)$.)

It can be shown (see [1] Theorem 1.1.4) that the essential spectrum of $-\Delta_\tau$ equals $[0,\infty)$ for every $\tau \in (-\pi,\pi]$. If $\tau \in [\frac{1}{2}\pi,\pi]$, the point spectrum is empty. Otherwise $-\Delta_\tau$ has exactly one negative simple eigenvalue $\lambda_0 = -\left(\frac{\cos(\frac{1}{4}\pi+\frac{1}{2}\tau)}{\cos\frac{1}{2}\tau}\right)^2$ with a positive eigenfunction $\varphi_{-\lambda_0} = \frac{e^{-\sqrt{-\lambda_0}|x|}}{4\pi|x|}$. As we pointed out in the introduction, we are interested in the case $-\Delta_\tau$ has a negative eigenvalue. So we assume $\tau \in (-\pi, \frac{1}{2}\pi)$ in the sequel.

Using the so-called Fermi pseudopotential (see [1], [5]), point interactions can also be denoted as follows. First we consider the closure of $-\Delta$ in $H^{-2,2}(\mathbb{R}^3)$

$$-\Delta_{-2,2} : L^2(\mathbb{R}^3) \subseteq H^{-2,2}(\mathbb{R}^3) \longrightarrow H^{-2,2}(\mathbb{R}^3).$$

This operator will be perturbed by an extended Dirac measure, which is not only defined on the (continuous) elements of $H^{2,2}(\mathbb{R}^3)$, but also on φ_+ and φ_-. We define this extension in the following way

$$4\pi\delta\frac{\partial}{\partial r}r : H^{2,2}(\mathbb{R}^3) \oplus [\varphi_+ + e^{i\tau}\varphi_-] \longrightarrow H^{-2,2}(\mathbb{R}^3)$$

and for $u \in H^{2,2}(\mathbb{R}^3)$ and $c \in \mathbb{C}$

$$\begin{aligned} & 4\pi\delta\frac{\partial}{\partial r}r(u + c(\varphi_+ + e^{i\tau}\varphi_-)) := \\ = \;& 4\pi\left[\frac{\partial}{\partial r}r(u + c(\varphi_+ + e^{i\tau}\varphi_-))\right](0).\delta_0 = \\ = \;& \left[4\pi u(0) + c(-\frac{1}{2}\sqrt{2} + \frac{1}{2}\sqrt{2}i) + ce^{i\tau}(-\frac{1}{2}\sqrt{2} - \frac{1}{2}\sqrt{2}i)\right].\delta_0. \end{aligned}$$

Definition 1 For $\alpha \in \mathbb{R}$ the part of $-\Delta_{-2,2} + 4\pi\alpha\delta\frac{\partial}{\partial r}r$ in $L^2(\mathbb{R}^3)$ is denoted by $-\Delta + 4\pi\alpha\delta\frac{\partial}{\partial r}r$.

By calculation and using the remarks made earlier, one can prove

Proposition 2 *Define for* $\tau \in (-\pi, \frac{1}{2}\pi)$ *the coefficient* $\alpha_\tau = \frac{\cos\frac{1}{2}\tau}{\cos(\frac{1}{4}\pi+\frac{1}{2}\tau)}$. *Then* $-\Delta_\tau$ *equals* $-\Delta + 4\pi\alpha_\tau\delta\frac{\partial}{\partial r}r$ *on* $D(-\Delta_\tau)$. *Moreover*

$$D(-\Delta_\tau) = \{u_0 + c\varphi_{\frac{1}{\alpha_\tau^2}} \ : \ u_0 \in H^{2,2},\ u_0(0) = 0,\ c \in \mathbb{C}\}$$

where $\varphi_{\frac{1}{\alpha_\tau^2}} = \frac{e^{-\frac{1}{\alpha_\tau}|x|}}{4\pi|x|}$ *is an eigenfunction of* $-\Delta_\tau$ *with eigenvalue* $-\frac{1}{\alpha_\tau^2}$. *(Note that* $\tau \to \alpha_\tau$ *defines a bijection from* $(-\pi, \frac{1}{2}\pi)$ *onto* $(0, \infty)$.*)*

Consider the following class of potentials $\mathcal{C}$

$$\mathcal{C} := \{V = V_1 + V_2 \in L^2 + L^\infty \ : \ V \geq 0,\ \lim_{|x|\to\infty} V_2(x) = 0,\ \frac{1}{r}V \in L^2_{loc}\}.$$

As $H^{2,2}(\mathbb{R}^3)$ can be imbedded into the space of continuous functions and $\varphi_{\frac{1}{\alpha^2}}$ has a singularity of order $\frac{1}{r}$, the perturbation of $-\Delta + 4\pi\alpha\delta\frac{\partial}{\partial r}r$ by V is well defined as an operator in $L^2(\mathbb{R}^3)$. It can be shown that for every $\alpha \in (0, \infty)$ the operator $(-\Delta + 4\pi\alpha\delta\frac{\partial}{\partial r}r) + V$ can also be obtained as selfadjoint extensions of $-\Delta + V$, restricted to $C_0^\infty(\mathbb{R}^3\backslash\{0\})$. Therefore we shall write $-\Delta + V + 4\pi\alpha\delta\frac{\partial}{\partial r}r$ instead of $(-\Delta + V) + 4\pi\alpha\delta\frac{\partial}{\partial r}r$ and call it a point interaction. In the following theorem we summarize some properties of point interactions.

Theorem 3 *(i) Let* $\alpha > 0$. *The essential spectrum of the point interaction* $-\Delta+4\pi\alpha\delta\frac{\partial}{\partial r}r$ *is* $[0, \infty)$. *There is exactly one eigenvalue* $-\frac{1}{\alpha^2}$, *which is simple, with corresponding eigenfunction* $\varphi_{\frac{1}{\alpha^2}} = \frac{e^{-\frac{1}{\alpha}|x|}}{4\pi|x|}$. *Moreover* $\varphi_{\frac{1}{\alpha^2}}$ *minimizes the functional*

$$f \longmapsto -\frac{1}{\alpha^2}\int_{\mathbb{R}^3} |f|^2 dx + \int_{\mathbb{R}^3} \nabla\varphi^{-1}_{\frac{1}{\alpha^2}}\bar{f}\ \nabla\varphi^{-1}_{\frac{1}{\alpha^2}}f\ \varphi^2_{\frac{1}{\alpha^2}}\,dx$$

under the condition $\|f\|_2 = \|\varphi_{\frac{1}{\alpha^2}}\|_2$.
(ii) Let $\alpha > 0$ *and* $V \in \mathcal{C}$. *The essential spectrum of the point interaction* $-\Delta + V + 4\pi\alpha\delta\frac{\partial}{\partial r}r$ *is* $[0, \infty)$. *This point interaction has at most one negative eigenvalue* λ_0, *which is simple, and* $\psi_{\lambda_0} := (\lambda_0 - \Delta + V)^{-1}_{-2,2}\delta_0$ *is an eigenfunction. This eigenfunction minimizes the functional*

$$f \longmapsto -\frac{1}{\alpha^2}\int_{\mathbb{R}^3} |f|^2 dx + \int_{\mathbb{R}^3} \nabla\varphi^{-1}_{\frac{1}{\alpha^2}}\bar{f}\ \nabla\varphi^{-1}_{\frac{1}{\alpha^2}}f\ \varphi^2_{\frac{1}{\alpha^2}}\,dx + \int_{\mathbb{R}^3} V|f|^2 dx$$

under the condition $\|f\|_2 = \|\psi_{\lambda_0}\|_2$. *Consequently* $\lambda_0 \geq -\frac{1}{\alpha^2}$.

Statement (i) is a consequence of Proposition 2 and the remarks made at the beginning of this section. See also [3] Chapters 3 and 9. For the proof of (ii) we refer to [3], Chapters 3 and 5. There, explicit expressions for the resolvents of point interactions are given. The statement on the essential spectrum and the fact that the point interaction has at most one negative eigenvalue, which is simple, can then be derived by using the Weinstein-Aronszajn determinant (see e.g. [6] IV 6).

3 The nonlinear problem

Let $V \in \mathcal{C}$ and $\alpha > 0$. We denote the operator $-\Delta + V + 4\pi\alpha\delta\frac{\partial}{\partial r}r$ in $L^2(\mathbb{R}^3)$ by B_α. Assume $\lambda_0 < 0$ is an eigenvalue of B_α with eigenfunction ψ_{λ_0}. Consider

$$B_\alpha u + |u|^{s-1}u = \lambda u \tag{3.5}$$

where $1 < s < \frac{3}{2}$, $\lambda \in \mathbb{R}$ and $u = u_0 + c\varphi_{\frac{1}{\alpha^2}} \in D(B_\alpha)$. (It follows from $1 < s < \frac{3}{2}$ that for $u \in D(B_\alpha)$ we have $|u|^{s-1}u \in L^2(\mathbb{R}^3)$.)

Let $(X, \|\ \|_X)$ denote $D(B_\alpha)$ supplied with the norm, defined by $\|u_0 + c(\varphi_{\frac{1}{\alpha^2}})\|_X := \|u_0\|_{2,2} + |c|$ for $u_0 + c(\varphi_{\frac{1}{\alpha^2}}) \in D(B_\alpha)$.

Definition 4 A solution (to problem (3.5)) is a pair $(\lambda, u) \in \mathbb{R} \times X$ satisfying (3.5). A solution (λ, u) is called positive if $u > 0$.

Theorem 5

1. *If* (λ, u) *is a nontrivial solution then* $\lambda > \lambda_0$.
2. *If* (λ, u) *is a nontrivial positive solution then* $\lambda \in (\lambda_0, 0]$.
3. *There is a constant* $M > 0$ *such that: if* (λ, u) *is a positive solution then* $\int_{\mathbb{R}^3} u\psi_{\lambda_0}dx \leq M|\lambda - \lambda_0|^{\frac{1}{s-1}}$ *and* $\int_{\mathbb{R}^3} |u|^{s+1}dx \leq M$ *and* $\int_{\mathbb{R}^3} |u|^2dx \leq \frac{M}{-\lambda}$.
4. *There is a* C^1*-function* $\Psi : (\lambda_0, 0) \longrightarrow X$ *such that* $(\lambda, \Psi(\lambda))$ *is a positive solution for* $\lambda \in (\lambda_0, 0)$ *and* $\Psi(\lambda) \to 0$ *in* X *as* $\lambda \downarrow \lambda_0$.

5. *For* $\lambda \in (\lambda_0, 0)$ *the problem (3.5) has exactly three solutions:* $(\lambda, \Psi(\lambda))$, $(\lambda, 0)$ *and* $(\lambda, -\Psi(\lambda))$.

Remark Point interactions are also well defined in $L^p(\mathbb{R}^3)$ for $\frac{3}{2} < p < 3$ as negative generators of analytic semigroups (see [2] and [3]). If in (1.2) $-\Delta_\tau$ is replaced by a point interaction in $L^p(\mathbb{R}^3)$ for suitable $\frac{3}{2} < p < 2$, then we can allow $1 < s < 2$.

4 Outline of the proof of the main result

1. If (λ, u) *is a nontrivial solution then* $\lambda > \lambda_0$.
2. If (λ, u) *is a nontrivial positive solution then* $\lambda \in (\lambda_0, 0)$.

Suppose $(\lambda, u) \in \mathbb{R} \times X$ is a nontrivial (positive) solution of (3.5). Then u is a (positive) eigenfunction with eigenvalue λ of the operator $B_\alpha + |u|^{s-1}$. Note that $|u|^{s-1} \in \mathcal{C}$ and therefore $|u|^{s-1} + V \in \mathcal{C}$. So u is a (positive) eigenfunction with eigenvalue λ of the operator

$$-\Delta + V + |u|^{s-1} + 4\pi\alpha\delta\frac{\partial}{\partial r}r$$

and we can apply Theorem 3 to conclude that $\lambda > -\frac{1}{\alpha^2}$ $(\lambda \in (-\frac{1}{\alpha^2}, 0))$. Comparing the quadratic forms of the operators $-\Delta + V + 4\pi\alpha\delta\frac{\partial}{\partial r}r$ and $-\Delta + V + |u|^{s-1} + 4\pi\delta\frac{\partial}{\partial r}r$ it can be shown that $\lambda \geq \lambda_0$.

3. There is a constant $M > 0$ *such that: if* (λ, u) *is a positive solution then* $\int_{\mathbb{R}^3} u\psi_{\lambda_0} dx \leq |\lambda - \lambda_0|^{\frac{1}{s-1}}$ *and* $\int_{\mathbb{R}^3} |u|^{s+1} dx \leq M$ *and* $\int_{\mathbb{R}^3} |u|^2 dx \leq M$.

If we multiply (3.5) by ψ_{λ_0} and integrate over $\mathbb{R}^3$, Jensen's inequality (see e.g. [9]) can be used to derive the desired estimates.

4. There is a C^1*-function* $\Psi : (\lambda_0, 0) \longrightarrow X$ *such that* $(\lambda, \Psi(\lambda))$ *is a positive solution for* $\lambda \in (\lambda_0, 0)$ *and* $\Psi(\lambda) \to 0$ *as* $\lambda \downarrow \lambda_0$.

Define $F : \mathbb{R} \times X \longrightarrow L^2(\mathbb{R}^3)$ by

$$F(\lambda, u) = B_\alpha u + |u|^{s-1}u - \lambda u.$$

Lemma 6 *The partial derivatives F_u, F_λ and $F_{\lambda u}$ exist and are continuous. Moreover for $(\lambda, u) \in I\!R \times X$ and $v \in X$*

$$\begin{aligned} F_u(\lambda, u)v &= B_\alpha v + s|u|^{s-1}v - \lambda v \\ F_\lambda(\lambda, u) &= -u \\ F_{\lambda u}(\lambda, u)v &= -v \end{aligned}$$

and $N(F_u)$ is spanned by ψ_{λ_0}.

(Note that F is not twice differentiable with respect to u as $1 < s < \frac{3}{2}$.) By applying Theorem 1.7 of [4] one can prove

Proposition 7 *There is a neighbourhood U of $(\lambda_0, 0)$ in $I\!R \times X$, an interval $(-\epsilon, \epsilon)$ and continuous functions $\lambda : (-\epsilon, \epsilon) \longrightarrow I\!R$ and $\chi : (-\epsilon, \epsilon) \longrightarrow H^{2,2}(I\!R^3)$ such that $\lambda(0) = -\frac{1}{\alpha^2}$, $\chi(0) = 0$ and*

$$F^{-1}(0) \cap U = \{(\lambda(s), s(\psi_{\lambda_0} + \chi(s))) \; : \; |s| < \epsilon\} \cup \{(s, 0) \; : \; (s, 0) \in U\}.$$

Moreover $\Psi(s) := s(\psi_{\lambda_0} + \chi(s))$ is positive (resp. negative) when $s > 0$ (resp. $s < 0$).

Using the implicit function theorem it can be shown that, given a nontrivial positive solution (λ, u), locally all solutions can be described by a C^1-curve $\Psi(\lambda)$. This is stated in

Proposition 8 *Let $(\lambda, u) \in (\lambda_0, 0) \times X$ be a nontrivial solution. Then there is a neighbourhood N of λ in $I\!R$ and a C^1-curve $\Psi : N \to X$ such that $\Psi(\lambda) = u$ and for $\mu \in N$: $(\mu, \Psi(\mu))$ is a solution of (3.5). Moreover there is a neighbourhood U of (λ, u) such that all solutions of (3.5) in U are of the form $(\mu, \Psi(\mu))$.*

So far we have proven the existence of a C^1-curve $\Psi : (\lambda_0, \bar{\lambda}) \to X$ where $\bar{\lambda} \in (\lambda_0, 0]$, such that $(\lambda, \Psi(\lambda))$ is a positive solution for $\lambda \in (\lambda_0, \bar{\lambda})$ and $\Psi(\lambda) \to 0$ as $\lambda \to \lambda_0$. It is a consequence of Proposition 8 and the following proposition that $\bar{\lambda}$ can be taken equal to 0.

Proposition 9 *Either $\lim_{\lambda \uparrow \bar{\lambda}} \Psi(\lambda)$ exists in X and $(\bar{\lambda}, \lim_{\lambda \uparrow \bar{\lambda}} \Psi(\lambda))$ is a solution, or $\bar{\lambda} = 0$.*

We give an outline of the proof. Suppose $\bar{\lambda} < 0$. Let λ_n be a sequence in $(\lambda_0, 0)$ such that $\lambda_n \uparrow \bar{\lambda}$ as $n \to \infty$, and define $u_n := \Psi(\lambda_n)$. It can be shown that $u_{n_1} \leq u_{n_2}$ if $n_1 \leq n_2$ so by *2.* and Beppo Levi's Theorem

$$u_n \to \bar{u} \text{ in } L^2(\mathbb{R}^3) \text{ and } u_n \to \bar{u} \text{ in } L^{s+1}(\mathbb{R}^3)$$

for $\bar{u}(x) := \sup_{n \in N} u_n(x)$. The latter statement can be rewritten as

$$|u_n|^s \to |\bar{u}|^s \text{ in } L^{1+\frac{1}{s}}.$$

Note that $1 + \frac{1}{s} < 2$ and $1 + \frac{1}{s} > 1 + \frac{2}{3} = \frac{5}{3} > \frac{3}{2}$.

For $\omega \geq 0$

$$\omega u_n + B_\alpha u_n = -|u_n|^s + (\lambda_n + \omega) u_n.$$

In [2], [3] it is shown that for ω large enough $(\omega + B_\alpha)^{-1}$ is bounded on $L^{1+\frac{1}{s}}(\mathbb{R}^3)$ and on $L^2(\mathbb{R}^3)$. So writing $u_{1,n} := (\omega + B_\alpha)^{-1}(-|u_n|^s)$ and $u_{2,n} := (\omega + B_\alpha)^{-1}((\lambda_n + \omega)u_n)$ and $\bar{u}_1 := (\omega + B_\alpha)^{-1}(-|\bar{u}|^s)$ and $\bar{u}_2 := (\omega + B_\alpha)^{-1}((\lambda_n + \omega)\bar{u})$ we have

$$u_{1,n} \to \bar{u}_1 \text{ in } H_0^{2,1+\frac{1}{s}} \oplus [\varphi_{\frac{1}{\alpha^2}}] \subseteq L^{1+\frac{1}{s}}(\mathbb{R}^3)$$

and

$$u_{2,n} \to \bar{u}_2 \text{ in } H_0^{2,2} \oplus [\varphi_{\frac{1}{\alpha^2}}] \subseteq L^2(\mathbb{R}^3).$$

Using this, one can show that, writing

$$u_n = u_{0,n} + c_n \psi_{\lambda_0}$$

and

$$\bar{u} = \bar{u}_0 + \bar{c}\psi_{\lambda_0}$$

with $u_{0,n} \in H^{2,2}(\mathbb{R}^3)$, $\bar{u}_0 \in H^{2,2}(\mathbb{R}^3) + H^{2,1+\frac{1}{s}}(\mathbb{R}^3)$ and $c_n, \bar{c} \in \mathbb{C}$, that $\|u_{0,n} - \bar{u}_0\|_\infty \to 0$ and $|c_n - \bar{c}| \to 0$. From this we deduce that $\bar{u} \in D(B_\alpha)$ and $\|u_n - \bar{u}\|_X \to 0$.

5. For $\lambda \in (\lambda_0, 0))$ (3.5) has exactly three solutions $(\lambda, \Psi(\lambda))$, $(\lambda, 0)$ and $(\lambda, -\Psi(\lambda))$.

So far we have established the existence of a C^1-curve $\Psi : (\lambda_0, 0) \to X$ such that $(\lambda, \Psi(\lambda))$ is a solution for all $\lambda \in (\lambda_0, 0)$ and $\Psi(\lambda) \to 0$ as $\lambda \to \lambda_0$ and

$\Psi(\lambda) > 0$. Clearly also $(\lambda, -\Psi(\lambda))$ and $(\lambda, 0)$ are solutions for $\lambda \in (\lambda_0, 0)$. Suppose (μ, v) is a nontrivial solution for some $\mu \in (\lambda_0, 0)$. Without loss of generality we assume $v > 0$. We will show that $v = \Psi(\mu)$. It follows from Proposition 8 and Proposition 9 that there is a $\bar{\mu} \in (\lambda_0, \mu)$ and a C^1-curve $\Phi : (\bar{\mu}, 0) \to X$ such that $(\lambda, \Phi(\lambda))$ is a solution for all $\lambda \in (\bar{\mu}, 0)$ and $\Phi(\mu) = v$ and $\Phi(\lambda) > 0$. Following the proof of Proposition 9 (in fact some steps can be simplified in this case) it can be shown that $\Phi(\bar{\mu}) := \lim_{\lambda \downarrow \bar{\mu}} \Phi(\lambda)$ exists and $(\bar{\mu}, \Phi(\bar{\mu}))$ is a solution. So either $\Phi(\bar{\mu}) > 0$ or $\Phi(\bar{\mu}) = 0$. If $\Phi(\bar{\mu}) > 0$, the curve Φ can be continued by Proposition 8 for $\mu < \bar{\mu}$ in some neighbourhood of $\bar{\mu}$. If $\Phi(\bar{\mu}) = 0$ then $\bar{\mu} = \lambda_0$ as λ_0 is the only point of bifurcation by Proposition 7. From Proposition 7 it follows that the curve Φ and Ψ are the same and consequently $\Psi(\mu) = v$.

References

[1] S. Albeverio, F. Gesztesy, R. Høegh-Krohn, H. Holden, Solvable Models in Quantum Mechanics, Springer, New York, 1988.

[2] W. Caspers, Ph. Clément, Point interactions in L^p, Delft report 91-97, 1991, to appear in Semigroup Forum.

[3] W. Caspers, On Point Interactions, thesis in progress, Delft University of Technology.

[4] M.G. Crandall, P.H. Rabinowitz, Bifurcation from simple eigenvalues, J. Func. Anal. 8 (1971), 321-340.

[5] E. Fermi, Sul moto dei neutroni nelle sostanze idrogenate, Ricerca Scientifica 7 (1936), 13-52. English translation in 'E. Fermi Collected papers', Vol. I, Italy, 1921-1938, University of Chicago Pres, Chicago-London, 1962, 980-1016.

[6] T. Kato, Perturbation Theory for Linear Operators, Springer-Verlag, Berlin-Heidelberg-New York, 1966.

[7] M. Reed, B. Simon, Methods of Modern Mathematical Physics II: Fourier Analysis, Self-Adjointness,Academic Press, 1972.

[8] W.-M. Ni, J. Serrin, Existence and nonexistence theorems for ground states of quasilimear partial differential equations. The anomalous case, University of Minnesota, Mathematics report 84-150.

[9] M. Reed, B. Simon, Methods of Modern Mathematical Physics IV: Analysis of Operators, Academic Press, 1978.

[10] B. Simon, Schrödinger semigroups, Bull.Amer.Math.Soc., 7 (1982), 447-526.

Semipositone problems*

Alfonso Castro and Ratnasingham Shivaji

1. Introduction.

By a semipositone problem we mean a semilinear equation where the nonlinearity is nondecreasing and negative at the origin. A typical example is the Dirichlet problem

$$\Delta u + \lambda f(u) = 0 \text{ in } \Omega, \;\; u = 0 \text{ on } \partial\Omega \tag{1.1}$$

where $\lambda \in (0, \infty)$ is a parameter, Ω is a smooth bounded region in $\mathbb{R}^n$, Δ is the Laplacian operator and $f : \mathbb{R} \to \mathbb{R}$ is a locally Lipschitzian monotonically increasing function such that

$$f(0) < 0 \tag{1.2}$$

and $f(\sigma) > 0$ for some $\sigma > 0$. Semipositone problems naturally arise in various studies. For example, consider the Rozenwig–McArthur equations in the analysis of competing species where "harvesting" takes place (see[1]). The study of positive solutions to (1.1) subject to (1.2), unlike the positone case ($f(0) \geq 0$), turns into a nontrivial question as $\phi = 0$ is not a subsolution, making the method of sub–super solutions difficult to apply. Semipositone problems, again unlike positone problems, give rise to the interesting phenomenon of symmetry breaking. The reader is referred to the work by Smoller & Wasserman (see [2]) for seminal work in this direction. The reader is referred to the work by Amman (see [3]), Laetch (see [4]), Keller–Cohen (see [5]), Brown–Ibrahim–Shivaji (see [6]), etc. for information on the positone case.

*Supported in part by NSF Grant DMS - 8905936

G. R. Goldstein and J. A. Goldstein (eds.),
Semigroups of Linear and Nonlinear Operations and Applications, 109–119.

The rest of the paper is divided into six sections: positive solutions when $n = 1$, positive radial solutions, sign changing solutions when $n = 1$, sign changing radial solutions, positive solutions in general bounded regions, and open problems.

2. Positive Solutions when n = 1.

Noting that solutions to

$$u'' + \lambda f(u) = 0 \tag{2.1}$$

are symmetric about its critical points, i.e., if $u'(a) = 0$ then $u(a+x) = u(a-x)$, and satisfy

$$u' = \pm(C - 2\lambda F(u))^{\frac{1}{2}} \text{ where } F(s) = \int_0^s f(t)dt, \tag{2.2}$$

the following results can be proven. Let β, θ be the unique positive zeros of f, F respectively, and let $\Lambda = 2(\int_0^\theta (-F(s))^{-\frac{1}{2}} ds)^2$. Then:

Theorem 2.1. *Let f be convex and superlinear. Then the boundary value problem $u'' + \lambda f(u) = 0;\quad u(0) = 0 = u(1)$ has a unique positive solution for $\lambda \in (0, \Lambda]$ and no positive solutions for $\lambda > \Lambda$. Further, there exists a unique non-negative solution with $(n-1)$ interior zeros at $\lambda = n^2\Lambda$ for $n \in N$.*

Theorem 2.2. *Let f be concave, $\lim_{s\to\infty} f(s) = M, 0 < M \le \infty$ and $\lim_{s\to\infty} sf'(s) = 0$. Then there exists a $K \in (0, \Lambda)$ such that the boundary value problem $u'' + \lambda f(u) = 0;\quad u(0) = 0 = u(1)$ has no positive solutions for $\lambda \in (0, K)$, exactly two positive solutions for $\lambda \in (K, \Lambda]$, and a unique positive solution for $\lambda \in \{K\} \cup (\Lambda, \infty)$. Further, there exists a unique non-negative solution with $(n-1)$ interior zeros at $\lambda = n^2\Lambda$ for $n \in N$.*

We refer the reader to [7] for the proofs of these above results, further details, and for discussion in the case of concave–convex type nonlinearities.

Theorem 2.3. *Assume that the set* $S = (\pi^2 n^2/f'(\beta), \theta^2/[-2F(\beta)])$ *is non–empty where* $n \in N$. *Then the boundary value problem* $u'' + \lambda f(u) = 0; \quad u'(0) = 0 = u'(1)$ *has at least* $2n+1$ *positive solutions for each* $\lambda \in S$.

See [8] for the proof of Theorem 2.3 and further details.

Theorem 2.4. *Let* f *be convex and superlinear. Then there exists* $\Gamma(\alpha) \in (0, \Lambda)$ *such that the boundary value problem* $u'' + \lambda f(u) = 0;\ u(0) = 0 = u(1) + \alpha u'(1)$, *where* $\alpha \in (0, \infty)$, *has a positive solution for each* $\lambda \in (0, \Gamma(\alpha)]$, *and no positive solutions for* $\lambda > \Lambda$. *Further, there exists* $\Gamma_n(\alpha) \in ([2n-1]^2\Lambda/4, n^2\Lambda)$ *such that there exists unique non–negative solutions with* $(n-1)$ *interior zeros at* $\lambda = \Gamma_n(\alpha)$ *and at* $\lambda = n^2\Lambda$ *for* $n \in N$.

For the proof of Theorem 2.4, and further details including the evolution of these solutions as α varies, see [9].

We close this section by also referring to [10], where the evolution of the branch of positive solution for the boundary value problem $u'' + \lambda f(u) = 0;\ u(0) = 0 = u(1)$ as the nonlinearity f evolves from having one positive zero to three positive zeros is discussed.

3. Radial Positive Solutions.

Unlike in the one dimensional case, when Ω is a ball in $\mathbb{R}^n (n > 1)$, the problem (1.1) with Dirichlet boundary conditions has no non–negative solutions with interior zeros. Indeed in [11] the authors proved the following result which answered an open question posed in the celebrated paper of Gidas - Ni - Nirenberg (see [12]).

Theorem 3.1.

Let Ω *be a ball in* $\mathbb{R}^n (n > 1)$ *and* $u \geq 0$ *be a solution of the boundary value problem*

$\Delta u + \lambda f(u) = 0$; $x \in \Omega$, $u = 0$; $x \in \partial\Omega$. *Then* $u > 0$ *in* Ω *and hence, radially symmetric.*

On the structure of positive branch of radial solutions to the boundary value problem $\Delta u + \lambda f(u) = 0$; $x \in \Omega$, $u = 0$; $x \in \partial\Omega$ when Ω is a ball in $\mathbb{R}^n (n > 1)$ the following results are known.

Theorem 3.2. *Let* f *be convex, superlinear, and subcritical (i.e. there exists* A *and* $p \in (0, [n+2]/[n-2])$ *such that* $|f(u)| \leq A(1 + u^p)$. *Then there exists a unique positive solution for* λ *small and no positive solution for* λ *large.*

The reader is referred to [13]–[16] for the proofs of these results.

Theorem 3.3. *Let* f *be concave and sublinear. Then there exists* $\delta_1 < \delta_2$ *such that for* $\lambda \in (0, \delta_1)$ *there are no positive solutions, for* $\lambda \in (\delta_1, \infty)$ *there exists a unique stable positive solution, and for* $\lambda \in [\delta_1, \delta_2]$ *there exists a unique unstable positive solution.*

For the proof of these results see [17]–[18].

4. Sign Changing Solutions When n = 1.

Like in the case of positive solutions when $n = 1$, using the quadrature method a complete study of sign changing solutions can be achieved. In particular, see [19] where the authors establish:

Theorem 4.1. *Let* f *be superlinear,* $f(s)$ *convex for* $s > 0$ *and concave for* $s < 0$, *and* $\lim_{s \to -\infty}(f(s) - sf'(s)) > 0$. *Then for each* $\lambda \in (0, n^2\Lambda)$ *the boundary value problem* $u'' + \lambda f(u) = 0$; $u(0) = 0 = u(1)$ *has at least one solution* u *with* $2n$ *interior zeros with* $u'(0) < 0$, *at least one solution* u *with* $2n - 1$ *interior zeros with* $u'(0) < 0$, *at least one* *solution u with $2n-2$ interior zeros with $u'(0) > 0$, and at least one solution with $2n-1$ interior zeros with $u'(0) > 0$. Further there exists a $\lambda^* \in (0, \Lambda]$ independent of n such that for $\lambda \in (0, \lambda^*)$ the above type of solutions are unique.*

Also see [20] where sign changing solutions for the boundary value problem $u'' + \lambda f(u) = 0; u(0) = 0 = u(1) + \alpha u'(1)$ are discussed in detail. In particular the following existence result is proven.

Theorem 4.2.

Let f be super linear. Then there exists two sequences of nontrivial bifurcation points $\{(\lambda_n(\alpha), \theta)\}_{n=1}^{\infty}$, $\{(\sigma_n, \theta)\}_{n=1}^{\infty}$ with $\lambda_n(\alpha) < \sigma_n$ and $\theta > 0$ is such that $F(\theta) = 0$, where $F(s) = \int_0^s f(t)dt$, for the boundary value problem $u'' + \lambda f(u) = 0$, $u(0) = 0 = u(1) + \alpha u'(1)$, $\alpha \in (0, \infty)$. Further the branch of solutions (λ, u) where u has $(2n-2)$ interior zeros and $u'(0) > 0$ and the branch of solutions (λ, u) where u has $(2n-1)$ interior zeros and $u'(0) < 0$ bifurcates from $(\lambda_n(\alpha), \theta)$, while the branch of solutions (λ, u) where u has $(2n)$ interior zeros and $u'(0) < 0$ and the branch of solutions (λ, u) where u has $(2n-1)$ interior zeros and $u'(0) > 0$ bifurcates from (σ_n, θ).

See also [20] for the study on the evolution of the above branches as α varies

5. Sign Changing Radial Solutions.

Unlike positive solutions to (1.1), sign changing solutions in the case when Ω is a ball are not necessarily radially symmetric. However for radial solutions we have the following description of its branches. In what follows S denotes a connected component of $\{(\lambda, u) | \lambda > 0, u \in C(\bar{\Omega}), u \text{ is radial, and } (\lambda, u) \text{ satisfies } (1.1)\}$. Then we have:

Theorem 5.1.

For each nonempty S there exists a nonnegative integer k such that if $(\lambda, u) \in S$ then u has either $2k$ or $2k+1$ nodal hypersurfaces.

Theorem 5.2.

Suppose $(\lambda_0, u_0) \in S$. The function u_0 has $2k$ nodal hypersurfaces in Ω and $\nabla u_0(x) \neq 0$ for $x \in \partial\Omega$ iff there exists $(\lambda_1, u_1) \in S$ satisfying $u_0(0) = u_1(0)$ and u_1 has $2k+1$ nodal hypersurfaces in Ω. Here k is a nonnegative integer.

Theorem 5.3.

For any nonnegative integer k, there exists a unique unbounded branch of solutions $S = S_k$ (say) such that if $(\lambda, u) \in S_k$ then u has either $2k$ or $2k+1$ nodal hypersurfaces in Ω.

Theorem 5.4.

There exists $M > 0$ such that if (λ, u) is a solution to (1.1) with $\lambda < M$, then $(\lambda, u) \in S_k$ for some nonnegative integer k. Moreover $\nabla u(x) \neq 0$ for $x \in \partial\Omega$.

Theorem 5.5.

If f is convex and $f(t)/[tf'(t) - f(t)]$ is a nondecreasing function then for each $\lambda > 0$ the problem (1.1) has at most one positive solution u such that $(\lambda, u) \in S_0$.

The reader is referred to [15] for proofs of Theorems 5.1–5.5 and further details.

6. Positive Solutions in General Bounded Regions.

Suppose now that Ω is a smooth bounded region without any special symmetry. Then the following results are known to date.

Theorem 6.1.

Let f be a superlinear and assume that there exists $p \in (1, [n+2]/[n-2])$ and real numbers A and B such that $A(|u|^p - 1) \leq f(u) \leq B(|u|^p + 1)$. Then there exists $\sigma > 0$ such that for $\lambda < \sigma$ the problem (1.1) has a positive solution.

This and related results have been proven in [21]–[22]. An array of methods in nonlinear functional analysis such as the mountain pass lemma, Leray – Schauder degree, and a priori estimates for positive solutions were ingeniously utilized by the authors in proving these results.

Next in [23] the authors by nontrival usage of Green's identities prove the following theorem, which also holds in the case of Robin boundary conditions.

Theorem 6.2.

If f is convex then every non–negative solution of (1.1) is unstable.

The study in the case when f is concave has been achieved via the sub–super solution method. In particular, in [2], [24] and [25] the authors succeed in producing the much needed non–negative subsolution. In fact, in certain ranges of λ, an anti–maximum principle by Clement and Peletier (see [26]) was used in establishing this non–negative subsolution. The summary of the existence results in the f concave case is as follows:

Theorem 6.3. *If f is concave there exists $L < T$ such that for $\lambda > T$ the problem (1.1) has a positive solution, and no positive solutions for $\lambda < L$. Further for classes of concave nonlinearities (1.1) has a positive solution close to the smallest eigen value of $-\Delta$ with Dirichlet boundary condition.*

For related results in the case of Robin boundary conditions see [27].

7. Open Problems.

It seems that a rather significant analysis has been achieved in the case when (1.1) reduces to an ordinary differential equation. When this is not the case, many questions remain unanswered. In this section we summarize some of these important open questions.

(1) *Is there a bounded smooth region Ω in which (1.1) has a nonnegative solution with interior zeros?*

Note that by the maximum principle this is not possible in the case of positone problems. Also recall that even in the case of semipositone problems this is not possible when Ω is a ball in $\mathbb{R}^n; n > 1$. However, note that for semipositone problems such solutions exists when say $\Omega = \{(x, y)/0 < x < 1, \ -\infty < y < \infty\}$. This follows easily by recalling that such solutions exists when $n = 1$.

(2) *Given a general bounded region Ω, when f is convex and superlinear, it is true that there exists at most one positive solution? Also when f is concave and sublinear, is it true that there is a most one positive solution for λ large?*

(3) *Study of existence of multiple positive solutions for concave–convex type nonlinearities when $n > 1$.*

See [7] where multiplicity results were discussed in this situation when $n = 1$.

(4) *Study of sign changing solutions for general bounded regions Ω in $\mathbb{R}^n; n > 1$.*

(5) *Study of systems of equations which are of semipositone nature.*

REFERENCES

1. M. R. Myerscough, B. F. Gray, W. L. Hogarth, and J. Norbury, *An analysis of an ordinary differential equations model for a two species predator - prey system with harvesting and stocking*, J. Math. Biol. **30** (1992), 389–411.

2. J. Smoller and A. Wasserman, *Existence of positive solutions for semilinear elliptic equations in general domains*, Arch. Rational Mech. Anal. **98 (3)** (1987), 229–249.

3. H. Amman, *Fixed point equations and nonlinear eigenvalue problems in ordered Banach spaces*, SIAM Rev. **18** (1976), 620–709.

4. T. W. Laetch, *The number of solutions of a nonlinear two point boundary value problem*, Indiana Univ. Math. **J(20)** (1970/71), 1–13.

5. H. B. Keller and D. S. Cohen, *Some positone problems suggested by nonlinear heat generation*. J. Math. Mech. **16** (1976), 1361–1376.

6. K. J. Brown, M.M.A. Ibrahim, and R. Shivaji, *S-shaped bifuracation curves*, J. Nonlinear Analysis, TMA **5(5)** (1981), 475–486.

7. A. Castro and R. Shivaji, *Nonnegative solutions for a class of nonpositone problems*. Proc. Roy. Soc. Edin. **108(A)** (1988), 291–302.

8. A. Miciano and R. Shivaji, *Multiple positive solutions for a class of semipositone Neumann two point boundary value problems*, J. Math. Anal. Appl. (to appear).

9. V. Anuradha and R. Shivaji, *Nonnegative solutions for a class of superlinear multiparameter semipositone problems*. Preprint.

10. A. Khamayseh and R. Shivaji, *Evolution of bifurcation curves for semipositone problems when nonlinearities develop multiple zeroes*, J. Appl. Math. and Comp. (to appear).

11. A. Castro and R. Shivaji, *Nonnegative solutions to a semilinear Dirichlet problem in a ball are positive and radially symmetric*, Comm. in PDE **14(8 & 9)** (1989), 1091–1100.

12. B. Gidas, W. Ni, and L. Nirenberg, *Symmetry and related properties via the maximum principles*, Comm. Math. Phys. **68** (1979), 209–243.

13. A. Castro and R. Shivaji, *Nonnegative solutions for a class of radially symmetric nonpositone problems*, Proc. Amer. Math. Soc. **106(3)** (1989), 735–740.

14. K. J. Brown, A. Castro, and R. Shivaji, *Nonexistence of radially symmetric nonnegative solutions for a class of semipositone problems*, J. Diff. & Int. Eqn. **2(4)** (1989), 541–545.

15. A. Castro, S. Gadam, and R. Shivaji, *Branches of radial solutions for semipositone problems*, Preprint.

16. A. Castro and S. Gadam, *Nonexistence of bounded branches and uniqueness of positive solutions in semipositone problems*, Preprint.

17. I. Ali, A. Castro, and R. Shivaji, *Uniqueness and stability of nonnegative solutions for semipositone problems in a ball*, Proc. Amer. Math. Soc. (to appear).

18. A. Castro and S. Gadam, *Uniqueness of stable and unstable positive solutions for semipositone problems*, J. Nonlinear Analysis, TMA (to appear).

19. V. Anuradha and R. Shivaji, *Existence of infinitely many nontrivial bifurcation points*, Resultat der Mathematik (to appear).

20. V. Anuradha and R. Shivaji, *Sign changing solutions for a class of superlinear multiparameter semipositone problems*, Preprint.

21. S. Unsurangsie, *Existence of a solution for a wave equation and elliptic Dirichlet problems*, Ph.D. Thesis(1988), Univ. of North Texas.

22. W. Allegretto, P. Nistri, and P. Zecca, *Positive solutions of elliptic nonpositone problems*, J. Diff. & Int. Eqns (to appear).

23. K. J. Brown and R. Shivaji, *Instability of nonnegative solutions for a class of semipositone problems*, Proc. Amer. Math. Soc. **112** (1991), 121–124.

24. K. J. Brown and R. Shivaji, *Simple proofs of some results in perturbed bifurcation theory*, Proc. Roy. Soc. Edin. **93(A)** (1982), 71–82.

25. A. Castro, J. B. Garner and R. Shivaji, *Existence results for classes of sublinear semipositone problems*, Resultat der Mathematik (to appear).

26. Ph. Clement and L. A. Peletier, *An anti-maximum principle for second order elliptic operators*, J. Diff. Eqns. **34** (1979), 218–229.

27. W. Allegretto and P. Nistri, *Existence and stability for nonpositone elliptic problems*, J. Nonlinear

Analysis, TMA (to appear).

UNIVERSITY OF NORTH TEXAS DENTON, TX 76203

MISSISSIPPI STATE UNIVERSITY MISSISSIPPI STATE, MS 39762

A generation result for C-regularized semigroups

Ioana Cioranescu

Abstract. In this work we give a generation theorem for C-regularized semigroups which generalizes Oharu's results of [7]

Let X be a Banach space and C an injective bounded linear operator in X with dense range. According to Da Prato [4] (see also Davies and Pang [3]) we say that the family $\{S_t\}_{t\geq 0}$ of bounded linear operators in X is a C-regularized semigroup if the family is strongly continuous and

$$S_t S_s = CS_{t+s} \text{ for } t, s \geq 0, \ S_0 = C.$$

Exponentially bounded C-regularized semigroups were introduced and extensively studied in connection with the abstract Cauchy Problem. One of the most important result of the theory is the following Hille Yosida type theorem:

Let A be a linear, densely defined operator on X s.t. $(\omega, \infty) \subset \rho\ (A)$, for some $\omega \geq 0$; then the following are equivalent:

a) A generates a C-regularized semigroup $\{S_t\}_{t\geq 0}$ satifying $\|S_t\| \leq Me^{\omega t}$, for $t \geq 0$ and some $M > 0$.

b) C commutes with all the resolvents of A and

$$\left\|(\lambda-\omega)^n C(\lambda-A)^{-n}\right\| \leq M \text{ for } \lambda > \omega, n \in N$$

(see [3], [4], [5], [6], [9]).

G. R. Goldstein and J. A. Goldstein (eds.),
Semigroups of Linear and Nonlinear Operations and Applications, 121–128.

We note that de Laubenfels proved the above result without the hypothesis that C have dense range; moreover he initiated in [5] the study of C-regularized semigroups that may not be exponentially bounded. We present here a generation result for such C-regularized semigroups which generalizes the following result of Oharu (Theorem 2.3 of [8]): for a linear densely defined operator A such that $(\omega, \infty) \subset \rho(A)$, the following condition is sufficient in order to generate a continuously differentiable semigroup of operators on $D(A^k)$:
for every $t \geq 0$ there is a $M(t) > 0$ such that for $\lambda > \omega$, $0 \leq \frac{n}{\lambda} \leq t$, $n \in N$, $x \in D(A^k)$

$$\left\| \lambda^n R(\lambda;\ A)^n x \right\| \leq M(t) \| x \|_k, \text{ where } \| x \|_k = \sum_{j=1}^{n} \left\| A^j x \right\|.$$

Throughout this work C is linear, bounded, injective with dense range C(X), A has the domain $D(A) \subset X$, $\rho(A)$ is the resolvent set of A, $R(\lambda;\ A)$ the resolvent function at $\lambda \in \rho(A)$.

Our main result is:

Main Theorem: *Let A be a densely defined linear operator satisfying the conditions:*

i) there exists $\omega > 0$ so that $(\omega, \infty) \subset \rho(A)$ and $R(\lambda;\ A)\, C = C\, R(\lambda;\ A)$ for $\lambda > \omega$;

ii) for every $t > 0$ there exist a $M(t) > 0$ such that

$$\left\| \lambda^n C R(\lambda; A)^n \right\| \leq M(t) \quad \text{for } \lambda > \omega \text{ and } 0 \leq \frac{n}{\lambda} \leq t,\ n \in N.$$

Then A generates a C-regularized semigroup which is $O(M(t))$ as $t \to \infty$.

Proof: We denote $T_h = (I - hA)^{-1}$; then the condition ii) can be written as:

for every $t > 0$ there is a $M(t) > 0$ satisfying

$$\left\|CJ_h^n\right\| \le M(t) \text{ for } h^{-1} > \omega \text{ and } 0 \le nh \le t. \tag{1}$$

We shall prove that $S_t x = \lim_{h\to 0_+} C\,J_h^{[t/h]}$ exists for every $x \in X$, uniformly for t in finite intervals of R_+, and that $\{S_t\}_{t\ge 0}$ is a C-regularized semigroup of generator A with $\|S_t\| = O(M(t))$. We note that

$$C(D(A)) \subset \overline{C\left(D\left(A^2\right)\right)}. \tag{2}$$

Indeed, for $x = Cy$, $y \in D(A)$, we have

$$\left\|J_h x - x\right\| = \left\|hJ_h ACy\right\| = h\left\|CJ_h Ay\right\| \xrightarrow[h\to 0_+]{} 0.$$

Hence $x = \lim_{h\to 0_+} J_h x$, where $J_h x = CJ_h y \in C\left(D\left(A^2\right)\right)$.

Moreover it is an easy matter to see that $\overline{C(D(A))} = X$, also using (2) we obtain

$$\overline{C\left(D\left(A^2\right)\right)} = X. \tag{3}$$

We can now adapt Oharu's method (see [7]) to our case.

Let $x \in D(A^2), t_0 > 0, t \in [0, t_0]$ and $\ell^{-1} \ge h^{-1} > \omega$; then we have

$$\begin{aligned} \left\|C^2 J_\ell^{[t/\ell]} x - C^2 J_h^{[t/h]} x\right\| \le \left\|C^2 J_\ell^{[t/\ell]} x - C^2 J_\ell^{nk} x\right\| + \\ \left\|C^2 J_\ell^{nk} x - C^2 J_h^{k} x\right\| + \left\|C^2 J_h^{k} x - C^2 J_h^{[t/h]} x\right\|. \end{aligned} \tag{4}$$

Let us take $\ell = 2^{-j}$, $h = 2^{-m}$, $n = 2^{j-m}$ and $k = [2^m t]$, where $j, m \in N$ and $j \ge m$; then $h - n\ell = 0$ and $kh \in [0, t_0]$.

Using (1) we have

$$\left\| C^2 J_\ell^{[t/\ell]}x - C^2 J_\ell^{nk}x \right\| = \ell \left\| \sum_{p=1}^{r} C^2 J_\ell^{nk+p} Ax \right\| \le t_0 \|C\| \, M \, \|Ax\| \le h \|C\| \; M(t_0) \, \|Ax\|$$

where $0 \le r = [t/\ell] - nk \le n$.

Since $J_h^k x = J_h^{[t/h]}$, in order to estimate (4) we only have to calculate $\left\| C^2 J_\ell^{nk}x - C^2 J_h^k x \right\|$. We note that

$$J_\ell^{nk}x - J_h^k x = -\sum_{i=0}^{k-1}\left(J_h^{k-i} J_\ell^{ni} x - J_h^{k-i-1} J_\ell^{n(i+1)} x \right).$$

But we have

$$J_h^{k-i} J_\ell^{ni} x - J_h^{k-i-1} J_\ell^{n(i+1)} = h J_h^{k-i} J_\ell^{n(i+1)} Ax - J_h^{k-i}\left(J_\ell^{n(i+1)} x - J_\ell^{ni} x \right)$$

$$= (h - n\ell) J_h^{k-i} J_\ell^{n(i+1)} Ax + J_h^{k-i}\left(n\ell J_\ell^{n(i+1)} x - \ell \sum_{p=1}^{n} J_\ell^{ni+p} Ax \right) = \ell^2 J_h^{k-i} \sum_{p=1}^{n-1} \sum_{q=p}^{n-1} J_\ell^{ni+q+1} A^2 x$$

As $0 \le (k - i)\, h \le kh \le t_0$ and $0 \le (ni + q + 1)\ell \le k\ell n \le t_0$ we can use (1) to get

$$\left\| C^2 J_\ell^{nk} x - C^2 J_\ell^{k} x \right\| \le \ell^2 \sum_{i=0}^{k-1} \left\| C J_h^{k-i} \right\| \sum_{p=1}^{n-1} \sum_{q=p}^{n-1} \left\| C J_\ell^{ni+q+1} A^2 x \right\|$$

$$\le k n^2 \ell^2 M^2(t_0) \left\| A^2 x \right\| \le h t_0 M^2(t_0) \left\| A^2 x \right\|.$$

Then, for $x \in D(A^2), t_0 > 0, \ell = 2^{-j}$ and $h = 2^{-m}$, with $2^j \ge 2^m > \omega$, we have

$$\left\| C^2 J_\ell^{[t/\ell]} x - C^2 J_h^{[t/h]} x \right\| \le \text{const}\,(x, t_0) \cdot h.$$

It follows that $\lim_{h=2^m \to 0} C^2 J_h^{[t/h]}x$ exists uniformly for $t \in [0, t_0]$. Then by (3) we have that

$$S_t x = \lim_{h=2^{-m} \to 0} CT_h^{[t/h]}x, \quad t > 0 \tag{5}$$

exists for every $x \in X$, uniformly on finite subintervals of R_+. It is also clear that

$$\|S_t x\| \le M(t_0) \cdot \|x\|, \quad t \in [0, t_0], \quad x \in X. \tag{6}$$

We shall prove that $\{S_t\}_{t \ge 0}$ defined by (5) is a C-regularized semigroup whose generator is A. Consider $x \in D(A)$ and $t > 0$; then the functions $s \to CJ_h^{[s/h]}Ax$ are step functions on [0, t], uniformly bounded with respect with $s \in [0, t]$ and $h = 2^{-m}$, $m \in N$.

Since

$$J_h^{[t/h]}x - x = h \sum_{k=0}^{[t/h]-1} J_h^k Ax - hAx + hJ_h^{[t/h]}Ax$$

then

$$CJ_h^{[t/h]}x - Cx = \int_0^t CJ_h^{[s/h]}Ax\, ds - \int_{[t/h]h}^t CJ_h^{[s/h]}Ax\, ds - hAx + hT_h^{[t/h]}Ax.$$

Letting $h = 2^{-m} \to 0$, we obtain

$$S_t x - Cx = \int_0^t S_s Ax\, ds = A\int_0^t S_s x\, ds. \tag{7}$$

In particular it follows that $t \to S_t x$ is continuous on $[0, \infty)$, for every $x \in D(A)$; then the density of D(A) and (6) yields the strong continuity of the operator family $\{S_t\}_{t\geq 0}$. We can now use Theorem 2.6 in [5] to conclude that $\{S_t\}_{t\geq 0}$ is a C-regularized semigroup generated by an extension $\tilde{A}$ of A. We shall finally prove that $\tilde{A}$ = A. Indeed, we note that by the closedness of A and (6), we obtain from (7)

$$\int_0^t S_s x\, ds \in D(A) \text{ and } S_t x - Cx = A\int_0^t S_s x\, ds, \quad x \in X. \tag{8}$$

Let $x \in D(\tilde{A})$; then by (8)

$$A\left(t^{-1}\int_0^t S_s\, x ds\right) = \frac{S_t x - Cx}{t} \xrightarrow[t\to 0_+]{} C\tilde{A}x.$$

Since $\lim_{t\to 0_+} t^{-1}\int_0^t S_s x ds = Cx$ and A is closed, we get

$$Cx \in D(A) \text{ and } ACx = C\tilde{A}x, \quad x \in D(\tilde{A}).$$

It follows that $CD(\tilde{A}) \subset D(T)$ and $\tilde{A}|CD(\tilde{A}) \subset A$. Since $CD(\tilde{A})$ is a core of $\tilde{A}$ (see Tanaka [9], Theorem 2.1) $\tilde{A} = \overline{\tilde{A}|CD(\tilde{A})} \subset A$. Therefore $A = \tilde{A}$, i.e. A is the generator of $\{S_t\}_{t\geq 0}$.

We note that one can prove as in [7] that the limit in (9) is independent of the chosen sequence, so that we have

Corollary 1. *Suppose A is as in the Main Theorem; then the C-regularized semigroup generated by A is given by*

$$S_t x = \lim_{h\to 0_+} (I - hS)^{-[t/n]} x \tag{9}$$

where the convergence is uniform in t on bounded intervals.

We also mention that Tanaka [9] proved that formula (9) is true for every exponentially bounded C-regularized semigroup.

Corollary 2. *Let A be a densely defined operator satisfying the conditions:*

a) there is $\omega > 0$ s.t. $(\omega, \infty) < \rho(A)$

b) for every $t > 0$ there is $M(t) > 0$ such that

$$\left\| \lambda^n R(\lambda; A)^n x \right\| \leq M(t) \| x \|_k,$$

for some $k \in N$, $x \in D(A^k)$, $\lambda > \omega$ *and* $0 \leq \frac{n}{\lambda} \leq t$, $n \in N$.

Then A generates an $R(\lambda; A)^k$-regularized semigroup which is $O(M(t))$, for all $\lambda > \omega$.

Proof: We only have to take in the above theorem $C = R(\lambda; A)^k$, $\lambda > \omega$.

Remark 1. This result is essentially Theorem 4.2 in Oharu's [7]. We note that condition b) is also necessary in order that an densely defined operator A with $(\omega, \infty) \subset \rho(A)$ be the generator of an $R(\lambda; A)^k$-regularized semigroup; see: the Main Theorem in Sanekata [8] and Corollary 4.5 in de Laubenfels [5]. We can not prove that condition ii) in our Main Theorem is also sufficient; however in the next remark we shall present a particular case in which this occurs.

Remark 2. In [2] Beals studied closed densely defined operators such that $R(\lambda; A)$ exists and is polynomially bounded when $\operatorname{Re} \lambda > C |\operatorname{Im}\lambda|^a$, for some $c > 0$ and $0 < a < 1$; he essentially proved that in this situation A generates a C-regularized semigroup $\{S_t\}_{t\geq 0}$ where $C = e^{-\varepsilon(-A)b}$, for $\varepsilon > 0$, $a \leq b < 1$. If we put $U_t = S_t C^{-1}$, $t \geq 0$, then

$$\|U_t x\| \le M(t)\|C^{-1}x\|, \quad x \in C(X), \text{ with } M(t) = \|S_t\|, \; t \ge 0.$$

We can further adapt Sanekata's proof in [8] to obtain that A satisfies the property ii) of the Main Theorem. For details and a more general case we refer to [1].

References

1. I. Cioranescu, ***Sur le problème de Cauchy au sens des ultradistributions***, C. R. Acad. Sci. Paris, t. 300 Serie 1, No. 7, 197-200, (1985).

2. R. Beals, ***Semigroups and abstract Gevrey spaces***, J. Funct. Anal., 10, 300-308, (1972).

3. E. B. Davies and M. M. Pang, ***The Cauchy problem and a generalization of the Hille-Yosida theorem***, Proc. London Math. Soc. 55, 181-208, (1987).

4. G. Da Prato, ***Semigruppi regolarizzabili***, Ricerche Mat. 15, 223-248, (1966).

5. R. de Laubenfels, ***C-Semigroups and the Cauchy Problem***, J. Funct. Anal., to appear.

6. I. Miyadera, ***On the generators of exponentially bounded C-semigroups***, Proc. Japan Acad., 62, Ser. A, 239-242, (1986).

7. S. Oharu, ***Semigroups of linear operators in a Banach space***, Publ. RIMS, Kyoto Univ., 7, (2), 205-260, (1971).

8. N. Sanekata, ***Some remarks on the ACP***, Publ, RIMS, Kyoto Univ. 11, 51-65, (1975).

9. N. Tanaka, **On the exponentially bounded C-semigroups**, Tokyo J. Math., Vol. 10, No. 1, 107-117, (1987).

University of Puerto Rico
Department of Mathematics
Faculty of Natural Sciences
Box 23355
Río Piedras, Puerto Rico 00931

Smoothing properties of heat semigroups in infinite dimensions

Giuseppe Da Prato[1]

1. Introduction

Let us consider first a finite-dimensional problem

$$\begin{cases} u_t(t,x) = \dfrac{1}{2}\displaystyle\sum_{k=1}^{n} \lambda_k \frac{\partial^2 u(t,x)}{\partial x_k^2} \quad t \geq 0 \\ \\ u(0,x) = \varphi(x),\ x = (x_1, ..., x_n) \end{cases} \tag{1}$$

where $\varphi \in C_b(\mathbb{R}^n)$, the space of all uniformly continuous and bounded mappings $\mathbb{R}^n \to \mathbb{R}$ and $\lambda_1, ..., \lambda_n$ are positive numbers.

As well known, there exists a unique classical solution to (1), that can be represented by

$$u(t,\cdot) = S_n(t)\varphi, \quad \varphi \in C_b(\mathbb{R}^n), \tag{2}$$

where

$$S_n(t)\varphi = \prod_{k=1}^{n} T_k(t)\varphi, \quad \varphi \in C_b(\mathbb{R}^n) \tag{3}$$

and

$$\begin{aligned} &T_k(t)\varphi(x_1, ..., x_n) \\ &= (\pi\lambda_k t)^{-1/2} \int_{-\infty}^{+\infty} e^{-\frac{(x_k-\xi)^2}{\lambda_k t}} \varphi(x_1, ..., x_{k-1}, \xi, x_{k+1}, ..., x_n)d\xi. \end{aligned} \tag{4}$$

Moreover $S_n(t)$ is an analytic semigroup on $C_b(\mathbb{R}^n)$. We are interested to the infinite-dimensional generalization of (1); more precisely we are given a separable Hilbert space H with a complete orthonormal system $\{e_k\}$ and a

[1]Partially supported by the Italian National Project MURST "Problemi nonlineari nell'Analisi ..."

G. R. Goldstein and J. A. Goldstein (eds.),
Semigroups of Linear and Nonlinear Operations and Applications, 129–141.

sequence of nonnegative numbers $\{\lambda_k\}$. Then we set $x_k =< x, e_k >$ and we consider the problem

$$\begin{cases} u_t(t,x) = \dfrac{1}{2}\sum\limits_{k=1}^{\infty} \lambda_k \dfrac{\partial^2 u(t,x)}{\partial x_k^2}, \quad t \geq 0, \\ \\ u(0,x) = \varphi(x), \ \varphi \in \overline{C}_b(H), \end{cases} \tag{5}$$

where $\frac{\partial}{\partial x_k}$ means the directional derivative in the direction of e_k and $\overline{C}_b(H)$ is a suitable closed subspace of $C_b(H)$, see §2 below. Obviously, existence for problem (5) is related to the existence of the limit

$$S(t)\varphi =: \lim_{n\to\infty} S_n(t)\varphi, \ \varphi \in \overline{C}_b(H). \tag{6}$$

As remarked in [2], a necessary condition for the limit in (6) exist for all $\varphi \in \overline{C}_b(H)$, is the following

$$\sum_{k=1}^{\infty} \lambda_k < +\infty. \tag{7}$$

This fact follows easily by computing $S_n(t)\varphi$ when $\varphi(x) = e^{-\frac{1}{2}|x|^2}$

$$(S_n(t)\varphi)(x) = \left[\prod_{k=1}^{n}(1+\lambda_k t)\right]^{-1/2} \cdot e^{-\frac{1}{2}\sum_{k=1}^{n}\frac{x_k^2}{1+\lambda_k t}} e^{-\frac{1}{2}\sum_{k=n+1}^{\infty} x_k^2}.$$

From now on we assume that (7) holds, and we introduce the positive self-adjoint nuclear operator Q by setting

$$Q = \sum_{k=1}^{\infty} \lambda_k e_k \otimes e_k.$$

Here we have used the notation

$$(x \otimes y)z = x < y, z >, \ x, y, z \in H.$$

Then we write problem (5) as

$$\begin{cases} u_t(t,x) = \dfrac{1}{2}\text{Tr}\,[Q u_{xx}(t,x)] \\ \\ u(0,x) = \varphi(x), \ \varphi \in \overline{C}_b(H), \end{cases} \tag{8}$$

where Tr denotes the trace and the subscript x denotes the Fréchet derivative with respect to x. In §2 we show, following [2], that if (7) holds then the limit in (6) exists for all $\varphi \in \overline{C}_b(H)$.

In §3 we consider the time-dependent problem

$$\begin{cases} u_t(t,x) = \frac{1}{2} \text{Tr } [Q(t)u_{xx}(t,x)], \ t \in]0,T] \\ u(0,x) = \varphi(x), \ \varphi \in \overline{C}_b(H), \end{cases} \tag{9}$$

where $Q(t)$ are positive self-adjoint nuclear nuclear operators for all $t \in]0,T]$, and $Q(\cdot)$ is continuous on $]0,T] = \{s \in \mathbf{R} : 0 < s \leq T\}$. If

$$\int_0^T \text{Tr } [Q(t)]dt < +\infty, \tag{10}$$

we prove that there exists an evolution operator $U(t,s)$ so that the solution of (9) is given by $u(t,\cdot) = U(t,0)\varphi$.

As shown in [2], problem (9) naturally arises when one studies

$$\begin{cases} z_t(t,x) = \frac{1}{2}\text{Tr } [Qz_{xx}(t,x)] + < Bx, z_x(t,x) >, \\ z(0,x) = \varphi(x), \ \varphi \in \overline{C}_b(H), \end{cases} \tag{11}$$

where $B : D(B) \subset H \to H$ is the infinitesimal generator of a strongly continuous semigroup e^{tB} in H and Q is a self-adjoint positive operator, not necessarily nuclear. In fact, setting $u(t,x) = z(t, e^{tB}x)$ and $Q(t) = e^{tB}Qe^{tB^*}$, problem (11) reduces formally to problem (9), which can be solved, under assumption (10). In this case we can associate to problem (11) a semigroup $G_B(t)$ defined by

$$(G_B(t)\varphi)(x) = (U(t,0)\varphi)(e^{tB}x), \ t \geq 0, \ x \in H, \ \varphi \in \overline{C}_b(H). \tag{12}$$

§4 is devoted to prove a smoothing result; namely $G_B(t)\varphi \in \overline{C}_b^1(H)$ for any $\varphi \in \overline{C}_b(H)$. This result was obtained in [4] by using probabilistic tools as the Cameron–Martin formula. Here we give a direct simple proof not involving probability. We remark that, when B is dissipative, a non probabilistic completely different proof was given in [2].

We remark finally that a probabilistic approach to problems (8) and (11), (and also to more general ones) is known from several years, see [3], [6], [7], [9], whereas a purely analitic approach is quite recent, see [1].

2. The heat semigroup

We are given a separable Hilbert space H, we denote by $\mathcal{L}_1^+(H)$ the set of all self-adjoint nonnegative nuclear operators on H and by $C_b^\infty(H)$ the set of all mappings from H into $\mathbf{R}$ that are Fréchet differentiable togheter with their derivatives of any order, all derivatives being uniformly continuous and bounded. For any mapping $\varphi : H \to \mathbf{R}$ we set

$$\|\varphi\|_0 = \sup_{x\in H} |\varphi(x)|.$$

Moreover, if φ is k times Fréchet differentiable we set

$$|\varphi|_k = \sup_{x\in H} |\varphi^k(x)|,$$

and

$$\|\varphi\|_k = \|\varphi\|_0 + \sum_{h=1}^{k} |\varphi|_h.$$

We shall denote by $\overline{C}_b(H)$ (resp. $\overline{C}_b^k(H)$) the closure of $C_b^\infty(H)$ with respect to the norm $\|\cdot\|_0$, (resp. $\|\cdot\|_k$.) We recall that the space $\overline{C}_b(H)$, introduced in [2], is not dense in $C_b(H)$, see [8], but it contains several functions of interest for the applications.

For any linear operator $Q \in \mathcal{L}_1^+(H)$ we denote by $\{e_k^Q\}$ a complete orthonormal system in H and by $\{\lambda_k^Q\}$ a sequence of nonnegative numbers such that

$$Qe_k^Q = \lambda_k^Q e_k^Q, \quad k = 1, 2, ...$$

Then, for any $n = 1, 2, ...$ we consider the analytic semigroups on $\overline{C}_b(H)$ defined as

$$(T_n^Q(t)\varphi)(x) = (\pi\lambda_n^Q t)^{-1/2} \int_{-\infty}^{+\infty} e^{-\frac{(<x,e_n^Q>-\xi)^2}{\lambda_n^Q t}} \varphi\left(x + (\xi - <x, e_n^Q>)e_n^Q\right) d\xi \tag{13}$$

and

$$S_n^Q(t)\varphi = \prod_{k=1}^{n} T_k^Q(t)\varphi, \quad \varphi \in \overline{C}_b(H). \tag{14}$$

We remark that the infinitesimal generator of the semigroup $T_n^Q(\cdot)$, $n = 1, 2, ...$, is the operator $(D_n^Q)^2$ where D_n^Q represents the directional derivative on the direction e_n^Q.

The main result of this section is the theorem below whose proof may be found in [2]; however we sketch it for the reader's convenience.

Theorem 1 *Let $Q \in \mathcal{L}_1^+(H)$; then the following statements hold*

(i) For all $\varphi \in \overline{C}_b(H)$, there exists the limit

$$S^Q(t)\varphi =: \lim_{n\to\infty} S_n^Q(t)\varphi, \text{ in } \overline{C}_b(H). \tag{15}$$

(ii) $S^Q(t)$ is a strongly continuous semigroup of contractions in $\overline{C}_b(H)$ and its infinitesimal generator $\mathcal{A}^Q$ is the closure of the linear operator $\mathcal{A}_0^Q$ defined as

$$\mathcal{A}_0^Q\varphi = \text{ Tr } [Q\varphi_{xx}], \ \forall\varphi \in \overline{C}_b^2(H), \tag{16}$$

(iii) For all positive integer $k, \overline{C}_b^k(H)$ is an invariant subspace for the semigroup $S^Q(\cdot)$, the restriction of $S^Q(\cdot)$ to $\overline{C}_b^k(H)$ is a strongly continuous semigroup of contractions in $\overline{C}_b^k(H)$ and its infinitesimal generator is the part of $\mathcal{A}^Q$ in $\overline{C}_b^k(H)$.

(iv) Let $\varphi \in \overline{C}_b^2(H)$ and set $u(t,x) = (S^Q(t)\varphi)(x)$, $x \in H, t \geq 0$. Then u is continuous on $[0,+\infty[\times H$, moreover $u(\cdot,x) \in C^1([0,+\infty[: \mathbf{R})$ for all $x \in H$, $u(t,\cdot) \in \overline{C}_b^2(H)$ for all $t \geq 0$ and

$$u_t(t,x) = \frac{1}{2} \text{ Tr } [Qu_{xx}(t,x)], \ t \geq 0, \ x \in H. \tag{17}$$

Proof — In the proof we omit all superscripts Q for simplicity. If $\varphi \in \overline{C}_b^2(H)$ we have

$$\begin{aligned} S_{n+1}(t)\varphi - S_n(t)\varphi &= \prod_{k=1}^{n} T_k(t)\left(T_{n+1}(t)\varphi - \varphi\right) \\ &= \prod_{k=1}^{n} \lambda_{n+1} T_k(t) \int_0^t T_{n+1}(s) D_{n+1}^2 \varphi \, ds. \end{aligned}$$

It follows

$$\begin{aligned}\|S_{n+1}(t)\varphi - S_n(t)\varphi\|_0 &\le \tfrac{1}{2}\lambda_{n+1}t\|D^2_{n+1}\varphi\|_0 \\ &\le \tfrac{1}{2}\lambda_{n+1}t\|\varphi\|_2,\end{aligned}$$

and so

$$\sum_{n=1}^{\infty}\|S_{n+1}(t)\varphi - S_n(t)\varphi\|_0 \le \frac{1}{2}t \operatorname{Tr} Q\|\varphi\|_2.$$

Conclusions (i) and (ii) now follow from the fact that $\overline{C}_b^2(H)$ is dense in $\overline{C}_b(H)$,the estimate

$$\|S_n(t)\varphi\|_0 \le \|\varphi\|_0, \ \forall\, \varphi \in \overline{C}_b(H)$$

and standard arguments, see [2] for details.

As far as (iii) is concerned, one has just to repeat the previous argument by replacing $\overline{C}_b(H)$ with $\overline{C}_b^k(H)$ and $\overline{C}_b^2(H)$ with $\overline{C}_b^{k+2}(H)$.

We prove finally (iv). If $\varphi \in \overline{C}_b^2(H)$ we have

$$\frac{d}{dt}S_n(t)\varphi = S_n(t)\left(\frac{1}{2}\sum_{k=1}^{n}\lambda_k D_k^2\varphi\right) = \frac{1}{2}\sum_{k=1}^{n}\lambda_k D_k^2(S_n(t)\varphi).$$

By (iii) it follows that $S(\cdot)\varphi$ is differentiable in $[0,+\infty[$ and equation (17) holds. ∎

We end this section by giving some results, which will be useful in §3, connecting different semigroups $S^Q(\cdot)$ and $S^R(\cdot)$, for $Q, R \in \mathcal{L}_1^+(H)$.

Proposition 2 *Let $Q, R \in \mathcal{L}_1^+(H)$; then the following statements hold*

(i) For all $t, s \ge 0$,

$$S^Q(t)S^R(s) = S^R(s)S^Q(t).$$

(ii) If $\varphi \in \overline{C}_b^2(H)$, then

$$S^Q(t)\varphi - S^R(t)\varphi = \int_0^t S^Q(t-s)S^R(s)(\mathcal{A}^Q\varphi - \mathcal{A}^R\varphi)ds. \qquad (18)$$

Proof — By recalling (13) we have

$$T_k^Q(t)T_k^R(s) = T_k^R(s)T_k^Q(t),$$

which clearly implies (i).

To prove (ii) set

$$u(t) = S^Q(t)\varphi,\ v(t) = S^R(t)\varphi,\ z(t) = u(t) - v(t),\ t \geq 0.$$

Then z is the solution to the initial value problem

$$\begin{cases} z'(t) = \mathcal{A}^Q u(t) - \left[\mathcal{A}^Q v(t) - \mathcal{A}^R v(t)\right] \\ \\ z(0) = \varphi, \end{cases}$$

and the conclusion follows from the variation of constants formula. ∎

Proposition 3 *Let $\{Q_n\}$ be an increasing sequence in $\mathcal{L}_1^+(H)$ strongly convergent to an element $Q \in \mathcal{L}_1^+(H)$. Then for any $\varphi \in \overline{C}_b(H)$ we have*

$$\lim_{n\to\infty} S^{Q_n}(t)\varphi = S^Q(t)\varphi, \text{ in } \overline{C}_b(H). \tag{19}$$

Proof– Since

$$\|S^{Q_n}(t)\varphi\|_0 \leq \|\varphi\|_0,\ \forall\, \varphi \in \overline{C}_b(H),$$

it suffices to prove (19) for $\varphi \in \overline{C}_b^2(H)$. In this case we have

$$\begin{aligned} \|\mathcal{A}^Q\varphi - \mathcal{A}^{Q_n}\varphi\|_0 &= \tfrac{1}{2}\|\text{Tr }[(Q - Q_n)\varphi_{xx}\|_0 \\ &\leq \tfrac{1}{2}\text{Tr }(Q - Q_n)\|\varphi\|_2 \\ &= \frac{1}{2}\|\varphi\|_2 \sum_{k=1}^{\infty} < (Q - Q_n)e_k^Q, e_k^Q > \to 0 \end{aligned}$$

as $n \to \infty$. Now by Proposition 2 we have

$$\|S^Q(t)\varphi - S^{Q_n}(t)\varphi\|_0 \leq t\|\mathcal{A}^Q\varphi - \mathcal{A}^{Q_n}\varphi\|_0$$

which implies (19) ∎

Proposition 4 *Let $Q, R \in \mathcal{L}^+(H)$; then we have*

$$S^Q(t)S^R(t) = S^{Q+R}(t). \tag{20}$$

Proof– Since

$$T_k^Q(t) = S^{\lambda_k^Q e_k^Q \otimes e_k^Q},$$

it follows

$$S_n^Q(t)S_n^R(t) = S^{\sum_{k=1}^n [\lambda_k^Q e_k^Q \otimes e_k^Q + \lambda_k^R e_k^R \otimes e_k^R]}(t).$$

Now, since

$$\lim_{n\to\infty} \sum_{k=1}^n [\lambda_k^Q e_k^Q \otimes e_k^Q + \lambda_k^R e_k^R \otimes e_k^R]x = (Q+R)x, \ \forall\, x \in H,$$

the conclusion follows from Proposition 3. ∎

3. Non autonomous problems

We are here concerned with the problem

$$\begin{cases} u_t(t,x) = \frac{1}{2}\,\mathrm{Tr}\,[Q(t)u_{xx}(t,x)],\ t \in [0,T], \\ u(0,\cdot) = \varphi \in \overline{C}_b(H), \end{cases} \tag{21}$$

under the following hypotheses

$$\begin{cases} (i)\ \ Q(t) \text{ is nuclear } \ \forall\, t \in]0,T]. \\ (ii)\ \ Q(\cdot)x \text{ is continuous in }]0,T]\ \forall\, x \in H. \\ (iii)\ \ \int_0^T \mathrm{Tr}\,[Q(t)]dt < +\infty. \end{cases} \tag{22}$$

We set

$$L(t,s)x = \int_s^t Q(\sigma)x d\sigma, T \geq t \geq s \geq 0,\ x \in H, \tag{23}$$

clearly $L(t,s) \in \mathcal{L}_1^+(H)$, for $T \geq t \geq s \geq 0$.

Theorem 5 *Assume that (22) holds and set*

$$U(t,s) = S^{L(t,s)}(1),\ T \geq t \geq s \geq 0. \tag{24}$$

Then the following statements hold

(i) For all $T \geq t \geq s \geq \sigma \geq 0$ *we have*

$$U(t,s)U(s,\sigma) = U(t,\sigma),\ U(s,s) = I. \tag{25}$$

(ii) $U(t,s)\varphi \in \overline{C}_b^2(H)$ *for all* $\varphi \in \overline{C}_b^2(H)$, $T \geq t \geq s \geq 0$.

(iii) $U(\cdot,s)\varphi \in C^1(]s,T];\mathbf{R})$ *and*

$$\frac{\partial}{\partial t}[U(t,s)\varphi] = \mathcal{A}^{Q(t)}U(t,s)\varphi,\ \forall\, t \in]s,T]. \tag{26}$$

Proof — (i) follows from Proposition 4, since

$$L(t,s) + L(s,\sigma) = L(t,\sigma)$$

for all $T \geq t \geq s \geq \sigma \geq 0$.

(ii) follows from Theorem 1–(iii).

Let us prove (iii). Let $\varphi \in \overline{C}_b^2(H)$ and $t, t+h \in]s,T]$, then, recalling (18) we have

$$\tfrac{1}{h}(U(t+h,s)\varphi - U(t,s)\varphi) = \tfrac{1}{h}(S^{L(t+h,s)}(1)\varphi - S^{L(t,s)}(1)\varphi)$$

$$= \int_0^1 S^{L(t+h,s)}(1-\rho)S^{L(t,s)}(\rho)\frac{1}{h}\operatorname{Tr}\left[\int_t^{t+h} Q(\sigma)d\sigma\varphi_{xx}\right]d\rho.$$

As h tends to 0 we have by Proposition 4,

$$\begin{aligned}\frac{\partial U(t,s)}{\partial t}\varphi &= \int_0^1 S^{L(t,s)}(1-\rho)S^{L(t,s)}(\rho)\operatorname{Tr}[Q(t)\varphi_{xx}]d\rho.\\ &= \mathcal{A}^{Q(t)}\varphi\end{aligned}$$

and (26) follows . ∎

4. Smoothing

4.1. Smoothing for equation (8)

Let $Q \in \mathcal{L}_1^+(H)$ and let $S^Q(t)$ be the semigroup defined by (15). It is useful to introduce, for any positive integer k, the semigroup

$$S_{(k)}^Q(t) = \prod_{n=1, n\neq k}^{\infty} T_n^Q(t). \tag{27}$$

We remark that the existence of the infinite product in (27) follows from Theorem 1, applied to the Hilbert space H_1 generated by all element of the basis $\{e_h^Q\}$ with the exception of e_k.

Theorem 6 *For all* $\varphi \in \overline{C}_b^1(H)$, $S^Q(t)\varphi$ *belongs to*$\overline{C}_b^1(H)$, *and the following estimate holds*

$$\|Q^{1/2}(S^Q(t)\varphi)_x\|_0 \leq t^{-1/2}\|\varphi\|_0. \tag{28}$$

Proof — Let $\varphi \in \overline{C}_b^1(H)$. We first remark that $S^Q(t)\varphi \in \overline{C}_b^1(H)$ by Theorem 1. Now setting $x_k =< x, e_k^Q >$, $k = 1, 2, ...$, we have

$$\frac{\partial}{\partial x_k}(S^Q(t)\varphi) = S_{(k)}^Q(t)T_k^Q(t)\varphi.$$

Since, as easily checked,

$$\left\|\frac{\partial}{\partial x_k}(T_k^Q(t)\varphi)\right\|_0 \leq \left(\lambda_k^Q t\right)^{-1/2}\|\varphi\|_0,$$

it follows

$$\left\|\frac{\partial}{\partial x_k}(S^Q(t)\varphi)\right\|_0 \leq \left(\lambda_k^Q t\right)^{-1/2}\|\varphi\|_0,$$

which is equivalent to

$$\left\|Q^{1/2}\frac{\partial}{\partial x_k}(S^Q(t)\varphi)e_k^Q\right\|_0^2 \leq \frac{1}{t}\|\varphi\|_0^2,$$

and now (28) follows immediately. ∎

Remark 7 By an easy density argument it follows, by Theorem 6, that $S^Q(t)\varphi$ possesses all directional derivatives, with respect to e_k^Q, $k = 1, 2, ..$ that, in addition, belong to $\overline{C}_b(H)$. Similar results can be obtained from higher order directional derivatives. ■

4.2. Smoothing for equation (11)

We are given a self–adjoint nonnegative operator Q, not necessarily nuclear, and a closed linear operator B, infinitesimal generator of a strongly continuous semigroup e^{tB}. Then we set

$$Q(t) = e^{tB}Qe^{tB^*}, L(t)x = L(t,0)x = \int_0^t Q(s)x ds, \; x \in H. \tag{29}$$

In this section we assume that $Q(t)$ fulfills (22) and study the semigroup G_B defined by (12), which can be written, recall Proposition 4, as

$$(G_P\varphi)(x) = (S^{L(t)}(1)\varphi)(e^{tB}x), \; \varphi \in \overline{C}_b(H), \; x \in H. \tag{30}$$

Remark that, if Q is nuclear, then (22) is automatically fulfilled.

We assume moreover

$$e^{tB}(H) \subset (L(t))^{1/2}(H), \; \forall\, t > 0. \tag{31}$$

As well known, hypothesis (31) is equivalent to the null controllability of the dynamical system

$$\xi'(t) = B\xi(t) + Q^{1/2}u(t), t \geq 0, \; \xi(0) = x \in H \tag{32}$$

in any time $t > 0$, where $\xi(t)$ is the state and $u(t)$ the control at time t, and it is always fulfilled if $Q = I$. In fact in this case, setting $u(s) = -\frac{1}{t}e^{sB}, s \in [0, t]$, one has $\xi(t) = 0$. For results on controllability see for instance [5].

If (31) holds we can define a linear bounded operator

$$\Gamma(t) = (L(t))^{-1/2}e^{tB}, \; t > 0, \tag{33}$$

where $(L(t))^{-1/2}$ denotes the pseudo–inverse of $(L(t))^{1/2}$.

Theorem 8 *Assume that* (22) *and* (31) *hold. Then, for any* $\varphi \in \overline{C}_b(H)$ *and any* $t > 0$, $G_B(t)\varphi$ *belongs to* $\overline{C}_b^1(H)$ *and*

$$\|(G_B(t)\varphi)_x\|_0 \leq \|\Gamma(t)\|\|\varphi\|_0. \tag{34}$$

Proof — Let first $\varphi \in \overline{C}_b^1(H)$. Then we have

$$\begin{aligned}(G_B(t)\varphi)_x &= e^{tB^*}(S^{L(t)}\varphi)_x(e^{tB}x) \\ &= \Gamma^*(t)L^{1/2}(t)(S^{L(t)}\varphi)_x(e^{tB}x).\end{aligned}$$

By Theorem 6 it follows

$$\|(G_B(t)\varphi)_x\|_0 \leq \|\Gamma^*(t)\|\|\varphi\|_0,$$

so (34) is proved for $\varphi \in \overline{C}_b^1(H)$; the general case follows from the density of $\overline{C}_b^1(H)$ in $\overline{C}_b(H)$. ∎

Remark 9 If $Q = I$ the following estimate can be proved, see [4],

$$\|\Gamma(t)\| \leq t^{-1/2} \sup_{s\in[0,t]} \|e^{tB}\|. \tag{35}$$

References

1. P. Cannarsa. G. Da Prato. *A semigroup approach to Kolmogoroff equations in Hilbert spaces*, Appl. Math. Lett. **4**, 49–52 (1991).

2. P. Cannarsa. G. Da Prato. *On a functional analysis approach to parabolic equations in infinite dimensions*, preprint Scuola Normale Superiore di Pisa No. 139. (1992).

3. Yu. Daleckii. S. Fomin. MEASURES AND DIFFERENTIAL EQUATIONS IN INFINITE DIMENSIONAL SPACES, Kluwer (1991).

4. G. Da Prato. J. Zabczyk. *Smoothing properties of transition semigroups in Hilbert spaces,* Stochastics and Stochastic Reports., 35, 63-77 (1991).

5. G. Da Prato. J. Zabczyk. STOCHASTIC EQUATIONS IN INFINITE DIMENSIONS, Encyclopedia of Mathematics and its Applications, Cambridge University Press (1992).

6. B. Gaveau. *Noyau de probabilité de transition de certains opérateurs d'Ornstein Uhlenbeck dans les espaces de Hilbert* C. R. Acad. Sc. Paris 293, 469-472 (1981).

7. L. Gross. *Potential Theory in Hilbert spaces* J. Funct. Analysis 1 , 123-189 (1965).

8. A. S. Nemirovski. S. M. Semenov. *The polynomial approximation of functions in Hilbert spaces*, Mat. Sb. (N.S), **92**, 134, 257-281 (1973).

9. M. A. Piech. *A fundamental solution of the parabolic equation in Hilbert space* J. Funct. Analysis 3 , 85- 114 (1969).

Scuola Normale Superiore
Piazza dei Cavalieri 6
56126, Pisa, Italy

References

Locally Stable Dynamics for Reaction Diffusion Systems

W.E. Fitzgibbon*
S.L Hollis
J.J. Morgan*

1. Introduction

We shall be concerned with the dynamics of solutions to systems of reaction diffusion equations. To be more precise we consider a weakly coupled semilinear parabolic system of the form:

$$\partial u/\partial t = D\Delta u + f(u) \qquad x \in \Omega,\ t > 0 \tag{1.1a}$$

$$\partial u/\partial n = 0 \qquad x \in \partial\Omega, t > 0 \tag{1.1b}$$

$$u(x,0) = u_0(x) \qquad x \in \Omega. \tag{1.1c}$$

Here the dependent variable $u = (u_1, \dots, u_m)^T$ is an m-dimensional vector and D is a diagonal matrix with strictly positive entries d_i, $i = 1$ to m, along the diagonal. With a certain abuse of notation we let Δ denote the vector Laplace operator and $\partial/\partial n$ represents a vector-valued Neumann boundary operator. We hope that it will not introduce confusion when we use the same symbols subsequently in the text for scalar equations. The nonlinear expression $f(\ ) = (f_1(\ ) \dots f_m(\ ))^T$ shall be called the reaction vector field. We shall require that

$$f_i(\ ) \in C^1(\mathbb{R}^m) \quad \text{for} \quad i = 1 \text{ to } m. \tag{1.2}$$

G. R. Goldstein and J. A. Goldstein (eds.),
Semigroups of Linear and Nonlinear Operations and Applications, 143–157.

The initial data is required to be continuous; i.e.

$$u_0(\ ) \in (C(\overline{\Omega}))^m. \tag{1.3}$$

Finally we stipulate that Ω is of class C^k and that $\partial\Omega$ lies locally on one side of Ω. Unless stated explicitly otherwise we shall assume the hypotheses delineated above hold through the course of this text.

Reaction diffusion systems arise in a variety of scientific and engineering contexts. Areas of application include chemical processing, population biology, semiconductor theory, contaminant transport, geological modelling, reservoir flow, combustion theory and oncology. The qualitative behavior of such systems can be difficult to understand but such an understanding can be crucial to a variety of scientific problems. We shall be concerned with the standard engineering problem of describing the flow of system (1.1a-c) near equilibrium. Equilibrium solutions of (1.1a-b) are known to satisfy weakly coupled semilinear elliptic systems of the form

$$-D\Delta w = f(w) \qquad x \in \Omega \tag{1.4a}$$

$$\partial w/\partial\eta = 0 \qquad x \in \partial\Omega. \tag{1.4b}$$

Spatially homogeneous solutions to (1.4a-b) are provided by zeros, or critical points, of the vector field. Thus if $z_0 \in \mathbf{R}^m$ is such that

$$f(z_0) = 0 \in \mathbf{R}^m$$

then z_0 is a solution to (1.4a-b). However, cf. [13], it is well known that spatially inhomogeneous solutions may also exist.

We shall follow the following organization for the remainder of this paper. The next section overviews the familiar notion of invariant rectangles for reaction diffusion equations and demonstrates the technique of truncating the system in the vicinity of stationary solutions. The third section introduces a subsidiary notion of stability of steady-state solutions and outlines a procedure which demonstrates that this subsidiary notion of stability in fact implies stability. In the fourth section we adapt Lyapunov methods to examine the stability of constant steady-states and we conclude with a section of examples.

2. Invariant Regions and Truncated Systems

The material overviewed in this section is well known in the literature and the reader is referred to [14] for an excellent and detailed exposition. A closed subset M is said to be a forward invariant region for (1.1a-c) if $u_0(x) \in M$ for $x \in \Omega$ implies $u(x,t) \in M$ for $x \in \Omega$, $t > 0$. If

$$M = I_1 \times \cdots \times I_m \tag{2.1}$$

where each I_i is a closed (possibly unbounded) interval and the vector field $f(\ )$ does not point out of M along ∂M then M is an invariant region for (1.1a-c). In most applications we are concerned with nonnegative quantities and therefore M will usually denote the positive orthant, $\mathbf{R}^m_+$.

Throughout the remainder of this paper we assume M is a forward invariant region for (1.1a-c). Suppose w is a solution to (1.4a-b) such that $w(x) \in M$ for all $x \in \Omega$. Let B be an m-dimensional cube such that $w(x) \in \operatorname{int} B$ for all $x \in \Omega$, and let $\eta = \inf_{x \in \Omega} \operatorname{dist}(w(x), B)$. If $\eta > 0$ rename B as $b_1(\eta, w)$ and let $b_2(\eta, w)$ be the m-cube concentric to $b_1(\eta, w)$ with twice the diameter. We set $M_1 = M \cap b_1(\eta, w)$ and $M_2 = M \cap b_2(\eta, w)$.

If we mollify the characteristic function of $b_1(\eta, w)$ we can produce a cutoff function $\varphi_{\eta,w} \in C^\infty(\mathbf{R}^m, [0,1])$ such that

$$\begin{aligned} \varphi_{\eta,w}(v) &= 1 \quad \text{for} \quad v \in b_1(\eta, w) \\ \varphi_{\eta,w}(v) &= 0 \quad \text{for} \quad v \in \mathbf{R}^m - b_2(\eta, w). \end{aligned} \tag{2.2}$$

We truncate the vector field by componentwise multiplication by $\varphi_{\eta,w}$; i.e. we define $f[\eta, w] = (f_i[\eta, w])_{i=1}^m$ by

$$f_i[\eta, w](u) = \varphi_{\eta,w}(u) f_i(u). \tag{2.3}$$

We may now consider the following truncation of (1.1a-c):

$$\partial v/\partial t = D\Delta v + f[\eta, w](v) \qquad x \in \Omega, t > 0 \tag{2.4a}$$

$$\partial v/\partial n = 0 \qquad x \in \partial\Omega, t > 0 \tag{2.4b}$$

$$v(x, 0) = \varphi_{\eta,w}(u_0) u_0(x) \qquad x \in \Omega. \tag{2.4c}$$

We have the following existence and containment result:

Lemma 2.5. *System (2.4a-c) has an unique classical solution on $\overline{\Omega} \times [0, \infty)$. Moreover;*

(i) $v(\ ,t) \in M_2$ for $t \geq 0$, and

(ii) if $v(\ ,t) \in M_1$ for $0 \leq t < T$ then

$$v(x,t) = u(x,t) \quad \text{for} \quad x \in \overline{\Omega},\ 0 \leq t < T,$$

where $u(\ )$ is the solution to (1.1a-c).

Proof. We observe that M_2 is a bounded invariant region for (2.4a-c) because the vector field $f[\eta, w]$ is identically zero exterior to $b_2[\eta, w]$, and hence does not point out of M. Therefore solutions to (2.4a-c) exist globally and remain confined to M_2 for all time. Classical parabolic uniqueness theory together with the observation that $f[\eta, w] = f$ on $b_1(\eta, w)$ guarantee the second assertion. ■

3. ∞-r Stability

We introduce a subsidiary notion of uniform stability and demonstrate that this notion may be used to guarantee uniform stability of steady-state solutions to (1.1a-c). Our analysis will involve the standard $L_p(\Omega)$ spaces with $p \geq 1$. However, we shall also want to consider $L_p(\Omega)$ spaces with $\|u\|_{p,\Omega} = [\int |u|^p dx]^{1/p}$ with $0 < p < 1$. We point out that if $0 < p < 1$, $\| \ \|_{p,\Omega}$ is not a norm. However, it may be used to define a complete metric space.

We now introduce our notation of uniform stability:

Definition 3.1. *Let $z_0 \in M$ be a critical point for the vector field $f = (f_i)_{i=1}^m$. Then z_0 is said to be stable with respect to M if for all $\varepsilon > 0$ there exists a $\delta > 0$ so that $u_0(x) \in M$ for all $x \in \Omega$ and $\|u_{0_i} - z_{0_i}\|_{\infty,\Omega} < \delta$ for $i = 1$ to m imply*

(i) A classical solution to (1.1a-c) exists on $\Omega \times [0, \infty)$, and

(ii) $\|u_i(\cdot, t) - z_{0_i}\|_{\infty,\Omega} < \varepsilon$ for $i = 1$ to m and $t > 0$.

A critical point $z_0 \in M$ which is stable with respect to M is said to be asymptotically stable with respect to M if there exists a $\delta > 0$ so that if $u_0(x) \in M$ for all $x \in \Omega$ with $\|u_{0_i} - z_{0_i}\|_{\infty,\Omega} < \delta$ for $i = 1$ to m, then we have

(iii) $\lim_{t \to \infty} \|u_i(\cdot, t) - z_{0_i}\|_{\infty,\Omega} = 0$ for $i = 1$ to m.

Our motivation for defining stability with respect to M is that we shall wish to consider critical points belonging to ∂M. We now introduce our notion of $\infty - r$ stability.

Definition 3.2. *Let $z_0 \in M$ be a critical point of the vector field $f = (f_i)_{i=1}^m$ and let $r > 0$. Then z_0 is said to be ∞-r stable with respect to M if for all $\varepsilon > 0$ there exists a $\delta > 0$ so that $u_0 \in M$ and $\|u_{0_i} - z_{0_i}\|_{\infty,\Omega} < \delta$ for $i = 1$ to m imply*

(i) a classical solution to (1.1a-c) exists on $\Omega \times [0, \infty)$, and

(ii) $\|u_i(\cdot, t) - z_{0_i}\|_{0,\infty} < \varepsilon$ for $i = 1$ to m and $t > 0$.

A critical point which is $\infty - r$ stable with respect to M is said to be asymptotically ∞-r stable with respect to M if there exists a $\delta > 0$ so that if $u_0(x) \in M$ for all $x \in \Omega$ with $\|u_i - z_{0_i}\|_{\infty,\Omega} < \delta$ for $i = 1$ to m, then we have

(iii) $\lim_{t\to\infty} \|u_i(\cdot, t) - z_{0_i}\|_{r,\infty} = 0$ for $i = 1$ to m.

We are now in a position to provide the following theorem which is the foundation of our development.

Theorem 3.3. *Let w be a solution to (1.4a-b) such that $w(x) \in M$ for all $x \in \Omega$. If $r > 0$ and $w(\ )$ is a ∞-r stable steady-state solution of (2.4a-c) with respect to M then w is a uniformly stable solution of (1.1a-c) with respect to M.*

Discussion of Proof. This theorem requires a lengthy bootstrapping argument and the reader is referred to [2] for details. First one uses the boundedness of solutions to (2.4a-c) to argue that ∞-r stability implies ∞-2 stability. The difficult part of the argument involves the bootstrapping of ∞-2 stability to uniform stability. Here one uses energy arguments in conjunction with the parabolic regularity estimates of Ladyzhenskaja, Solonnikov and Ural'ceva [7] and a fractional Sobolev embedding theorem appearing in Amann [1]. ∎

If one has a priori knowledge of the boundedness of solutions to (1.1a-c), then one may dispense with the truncated systems (2.4a-c) and directly conclude that ∞-r stability of steady-state solutions to (1.1a-c) implies uniform stability. Finally, we point out that only the boundedness of the steady-state is essential and the arguments of Theorem 3.3 can also be applied to determine the stability of other distinguished bounded solutions such as periodic solutions and almost periodic solutions.

4. Diffusively Convex Lyapunov Functionals

As an application of the foregoing theory we discuss the Lyapunov stability of critical points or zeros of the vector field $f(\ )$. It should be clear that the system of ordinary differential equations,

$$du/dt = f(u) \tag{4.1a}$$

$$u(0) = u_0 \tag{4.1b}$$

determines spatially homogeneous solutions to (1.1a-c). Critical points of f give steady-state solutions to (4.1a-c) and spatially homogeneous steady-state solutions to (1.1a-c).

The most common tool for analyzing the local stability of critical points for systems of ordinary differential equations of the form (1. a-b) is the principle of linearized stability. If all the eigenvalues of the derivative of f at z_0 have negative real part then z_0 is locally asymptotically stable. If on the other hand, some of the eigenvalues have negative real part and some have positive real part the critical point z_0 is unstable. These ideas carry over to the context of semilinear parabolic equations; see, e.g., [5]. In the case of nonhyperbolic critical points, however, linearization methods do not apply.

Questions of nonlinear stability are frequently resolved by Lyapunov's direct method. Roughly speaking a Lyapunov function V is a nonnegative functional which is defined and continuously differentiable in a neighborhood of a critical point z_0 and is uniquely minimized in that neighborhood by z_0. If

$$\dot{V}(u) = \partial V(u) f(u) \leq 0 \tag{4.2}$$

in this neighborhood, then it follows that z_0 is a stable critical point. Asymptotic stability can be deduced from conditions such as

$$\dot{V}(u) < -\alpha V(u) \tag{4.3}$$

for some $\alpha > 0$. In certain cases the existence of Lyapunov functionals satisfying (4.2) in a neighborhood of a critical points of a system of ordinary differential equations carries over to the context of the associated reaction-diffusion system. For this reason, we introduce the notion of D-diffusively convex Lyapunov functionals for reaction-diffusion systems.

Definition 4.4. *Let D be the matrix of diffusion coefficients for (1.1a) and suppose that M is a forward invariant set for (1.1a-c). If $z_0 \in M$ is a critical point of f we say that a nonnegative functional V is a D-diffusively convex Lyapunov functional around z_0 provided that the following conditions hold:*

(i) *There exists a $\xi > 0$ so that $V \in C^2(M \cap B_\xi(z_0), \mathbf{R}^+)$.*

(ii) *There exist constants $r > 0$ and $K > 0$ so that $V(u) \geq K \sum_{i=1}^{m} |u_i - z_{0_i}|^r$ for $u \in B_\xi(z_0) \cap M$.*

(iii) *$V(z_0) = 0$.*

(iv) *The matrix $D\partial^2 V(u)$ is nonnegative definite for $u \in B_\xi(z_0) \cap M$. (Here $\partial^2 V(u)$ is the Hessian matrix* *of* V*).*

(v) $\partial V(u)f(u) \leq 0$ *for* $u \in B_\xi(z_0) \cap M$.

We remark that conditions (i), (iii) and (v) are essentially those which define a Lyapunov functional for (1.1a-b) around z_0 and that conditions (ii) and (iv) represent an additional strengthening of the concept. If the functional V is separable; i.e.,

$$V(u) = \sum_{i=1}^{n} V_i(u_i), \tag{4.6}$$

then we may insure (iv) by assuming that $V_i''(u_i) \geq 0$. In general, however, convexity of V does not suffice for condition (iv). It is relatively straightforward to see that D-diffusively convex Lyapunov functionals guarantee the persistence of stability of critical points. We have the following theorem:

Theorem 4.6. *Let* $z_0 \in M$ *be a critical point for the vector field* f *and* M *be a forward invariant set for the semilinear parabolic system (1.1a-c). If there exists a local* D*-diffusively convex Lyapunov function* V *for* f *around* z_0*, then* z_0 *is a stable steady-state for (1.1a-c) with respect to* M*. Moreover, if* V *also satisfies (4.3), then* z_0 *is asymptotically stable with respect to* M*.*

Proof. We choose $\eta > 0$ so that the cube $C_{2\eta}(z_0) \subseteq B_\xi(z_0)$ and construct the truncated vector field $f[\eta, z_0]$ as in (2.2), (2.3). If $v_0(x) \in C_{2\eta}(z_0) \cap M$ for $x \in \Omega$ it is immediately verified that

$$\partial V(v(x,t))f(v(x,t)) = \partial V(v(x,t))f[\eta, z_0](v(x,t)) \leq 0. \tag{4.7}$$

If we multiply the ith component of (2.12a) by $\partial V(v)/\partial v_i$ we obtain

$$(\partial V(v)/\partial v_i)\partial v_i/\partial t = d_i(\partial V(v)/\partial v_i)\Delta v_i + (\partial V(v)/\partial v_i)f_i[\eta, z_0](v).$$

If we integrate this expression on the space-time cylinder and sum the components, we observe that

$$\int_\Omega V(v(x,t))dx = -\int_0^T \int_\Omega (\nabla v)^T D\partial^2 V(v)\nabla v dx dt$$
$$+\int_0^t \int_\Omega \partial V(v) f[\eta, z_0](v)dx + \int_\Omega V(v_0(x))dx.$$

Hence by virtue of conditions (iv) and (v) in Definition 4.4 we have

$$\int_\Omega V(v(x,t))dx \le \int_\Omega V(v_0(x))dx. \tag{4.8}$$

Using (4.8) and the coerciveness of V we get

$$K\left[\sum_{i=1}^m \|v_i(\cdot,t) - z_{0,i}\|_{r,\Omega}\right] \le \left[\int_\Omega V(v(x,t)dx\right]^{\frac{1}{r}} \tag{4.9}$$
$$\le \left[\int_\Omega V(v_0(x,t))dx\right]^{\frac{1}{r}}$$
$$\le \rho\left(\sum_{i=1}^m \|v_{0_i} - z_{0_i}\|_{\infty,\Omega}\right)$$

for some continuous ρ with $\rho(0) = 0$ and $\rho(s) > 0$ for $s > 0$. This will insure ∞-r stability with respect to M, and from Theorem 3.3 we may conclude that z_0 is stable with respect to M. Finally, in case (4.3) holds, one has

$$\int_\Omega V(v(x,t))dx \le e^{-\alpha t}\int_\Omega V(v_0(x))dx, \tag{4.10}$$

and from this follows the asymptotic stability assertion. ■

In view of Theorem 4.6 one can be naturally lead to the attempt of using D-diffusively convex Lyapunov functions to analyze the stability of spatially non-homogeneous steady-state solutions. The following simple proposition squashes this endeavor for large classes of dynamical systems.

Proposition 4.11. *Let M be a forward invariant set for (1.1a-c) and let $V(v) = \sum_{i=1}^{m} V_i(v_i)$ be a nonnegative separable function which satisfies the defining hypotheses of Definition 4.4 except possibly (ii), for all points of M. If $w = (w_1, \ldots, w_m)^T \in M$ is a solution to (1.4a-b) then the following are true:*

(i) If there exists $\alpha > 0$ so that $V_i''(v_i) > \alpha$ for all $v = (v_1, \ldots, v_m)^T \in M$, then $f(w) = 0$.

(ii) If $V(v) = \sum_{i=1}^{m} c_i v_i$, $\partial V(v) f(v) \leq 0$ and $M \subseteq \mathbf{R}_+^m$, then there exists a $k \geq 0$ so that $y(x) = \sum_{i=1}^{m} c_i d_i w_i(x) = k$ for all $x \in \Omega$; i.e., $w(x)$ belongs to a closed bounded subset of the hyperplane $\{v \mid \Sigma c_i d_i v_i = k\} \cap \mathbf{R}_+^m$. (4.11)

Proof. In the first case we multiply the ith component of (2.35a) by $V_i'(w_i)$ to obtain

$$-d_i V_i'(w_i) \Delta w_i = V_i'(w_i) f_i(w). \tag{4.12}$$

If we sum these terms and integrate on Ω we have

$$\sum_{i=1}^{m} d_i \int_\Omega V_i''(w_i) |\nabla w_i|^2 dx = \sum_{i=1}^{n} \int_\Omega V_i'(w_i) f_i(w) dx \leq 0. \tag{4.13}$$

Consequently,

$$\sum_{i=1}^{m} \alpha d_i \int_\Omega |\nabla w_i|^2 dx = 0, \tag{4.14}$$

and we may conclude that each w_i is a constant. Therefore, because $w = (w_1 \ldots w_n)^T$ is a solution to (1.4a-b), we must have $f_i(w) = 0$. If we follow the same train of reasoning for the second case we may observe that $-\Delta(\Sigma c_i d_i w_i) \leq 0$. The fact that M is required to lie in $\mathbf{R}_+^m$ implies that $\Sigma d_i w_i \geq 0$, and hence we conclude from maximum principles that $\nabla(\Sigma c_i d_i w_i)$ vanishes and

$\Sigma c_i d_i w_i(x) = k$ for some constant $k \geq 0$. Thus $w(x)$ lies in the hyperplane $\{v \mid \Sigma c_i d_i v_i = k\}$. Continuity of w implies that its range is closed and bounded. ■

We conclude this section with a discussion of an example arising from the theory of chemical reactions. Differential equations which describe the dispersion and reaction of m-chemical species are generally of the form

$$\partial u/\partial t = D\Delta u + f(u) \tag{4.15}$$

where the ith component of the dependent variable $u = (u_1, \ldots, u_m)^T$ represents the concentration density of the ith chemical species. The vector field $f = (f_i)_{i=1}^m$ is assumed to be in each component a polynomial function of the components of u and is intended to model the chemical reaction kinetics. Gröger, in his study of dissipative chemical reactions, [4], introduced the following hypothesis.

(G) There exists a vector $e = (e_1, \ldots, e_m)^T$ with each $e_i > 0$ so that $f(e) = 0$ and

$$\sum_{i=1}^{m} f_i(u) \log(u_i/e_i) \leq 0. \tag{4.16}$$

The quantity $\sum_{i=1}^{m} f_i(u) \log(u_i/e_i)$ is known to have the physical interpretation of being a suitably scaled rate of chemical dissipation, and work on the mathematical theory of reaction networks, [7], confirms that many nontrivial systems satisfy this hypothesis. If the chemical species are required to remain confined to a reaction vessel for all time the appropriate boundary conditions are given by

$$\partial u/\partial n = 0 \qquad (x,t) \in \partial\Omega \times (0,\infty). \tag{4.17}$$

Finally, a condition of the form

$$f_i(u) \geq 0 \quad \text{for} \quad u \in \mathbf{R}_+^m \text{ with } u_i = 0 \tag{4.18}$$

together with the maximum principle insures that $\mathbf{R}_+^m$ is a forward invariant set for (4.15). We have the following proposition:

Proposition 4.19. *We consider (4.15) together with the boundary conditions (4.17). If all the conditions describing a dissipative chemical reaction outlined above hold, then the steady-state $u = e$ is uniformly stable with respect to $\mathbf{R}_+^m$. Moreover, the elliptic system*

$$-D\Delta w = f(w) \qquad x \in \Omega \tag{4.19a}$$

$$\partial w/\partial n = 0 \qquad x \in \partial\Omega \tag{4.20b}$$

has no spatially inhomogeneous positive solutions.

Proof. We define

$$V(u) = \sum_{i=1}^{m} V_i(u) = \sum_{i=1}^{m} (u_i \log(u/e_i) - u_i + e_i) \tag{4.20}$$

and verify that all conditions of Definition 4.4 hold locally about e. Consequently, Theorem 3.4 implies that e is uniformly stable. An argument analogous to the one of Proposition 4.12 insures the nonexistence of positive spatially inhomogeneous steady-states. ■

We remark that [2] gives a similar analysis for the stability of constant steady-states of the general diffusive Lotka-Volterra system outlined in [9]. The reader will find additional use of the structure of D-diffusively convex Lyapunov functions in [3], [10] [11], [12].

References

1. H. Amann, Existence and regularity for semilinear parabolic evolution equations, *Annali Scoula Nosmale Superiore - Pisa*, Serie IV, Vol. IX 593-676 (1984).

2. W. Fitzgibbon, S.L. Hollis and J.J. Morgan, Stability and Lyapunov functions for reaction-diffusion systems, Preprint.

3. W. Fitzgibbon, J. Morgan and R. Sanders, Global existence and boundedness for a class of inhomogeneous parabolic equations, *J. Nonlinear Analysis*, to appear.

4. K. Gröger, On the existence of steady-states of certain reaction diffusion systems, *Arch. Rat. Mech. Anal.*, 1: 297-306 (1986).

5. D. Henry, *Geometric Theory of Semilinear Parabolic Equations*, Lecture Notes in Mathematics, 840, Springer-Verlag, Berlin, (1981).

6. S. Hollis, R. Martin and M. Pierre, Global existence and boundedness in reaction-diffusion systems, *SIAM J. Math. Anal.*, 18: 744-761 (1987).

7. F. Horn and R. Jackson, General mass action kinetics, *Arch. Rat. Mech. and Anal.*, 47: 81-116 (1972).

8. O. Ladyzhenskaja, V. Solonnikov and N. Ural'ceva, *Linear and Quasilinear Equations of Parabolic Type*, AMS Trans., Vol 23, Amer. Math. Soc., Providence, R.I., (1968).

9. A. Leung, *Systems of Nonlinear Partial Differential Equations*, Kluwer Academic Publ., Boston, (1989).

10. J. Morgan, Boundedness and decay results for reaction-diffusion systems, *SIAM J. Math. Anal.*, 21: 1172-1181 (1990).

11. J. Morgan, Global existence for reaction diffusion systems, *SIAM J. Math. Anal.* 20: 1128-1144 (1989).

12. J. Morgan, Global Existence for a class of quasilinear reaction-diffusion systems, Preprint.

13. J. Murray, *Mathematical Biology*, Springer-Verlag, Berlin (1989).

14. J. Smoller, *Shock Waves and Reaction Diffusion Equations*, Springer-Verlag, Berlin, (1984).

Department of Mathematics
University of Houston
Houston, Texas 77204-3476 USA

Department of Mathematics
Armstrong State College
Savannah, Georgia 31419-1997, USA

Department of Mathematics
Texas A&M University
College Station, Texas 77843-3368, USA

---

*These authors gratefully acknowledge the support of NSF Grants DMS 9207064 and DMS 9208046 respectively.

J. Morgan, Global ...

... Phys. Rev. ...

... Hugo ...

Global Dynamics of Singularly Perturbed Hodgkin-Huxley Equations

W. Fitzgibbon [1]
M. Parrott
Y. You

In their well-known work, Hodgkin and Huxley considered the following model for nerve impulse transmission across an axon:

$$\frac{\partial V}{\partial t} - \frac{\partial^2 V}{\partial x^2} = g_{Na} m^3 h \left(V_{Na} - V\right) + g_K n^4 \left(V_K - V\right) + g_L \left(V_L - V\right) \tag{1}$$

$$\frac{\partial m}{\partial t} = \left(m_\infty - m\right) / \tau_m \tag{2}$$

$$\frac{\partial h}{\partial t} = \left(h_\infty - h\right) / \tau_h \tag{3}$$

$$\frac{\partial n}{\partial t} = \left(n_\infty - n\right) / \tau_n \tag{4}$$

Here x represents the longitudinal distance along the axon, t is time; V is the electrical potential in the nerve; $m(V)$, $h(V)$, $n(V)$ are chemical concentrations of N_a, K and other (leekage) ions which are nonnegative and are nonlinear functions of V; g_{Na}, g_K, g_L are the maximum conductances of these ions, V_{Na}, V_K, V_L are the constant equilibrium potentials for these ions; m_∞, h_∞, n_∞ are the steady state values; and τ_m, τ_h, τ_n are relaxation times.

[1]Supported by NSF Grant No. DMS 9207064

G. R. Goldstein and J. A. Goldstein (eds.),
Semigroups of Linear and Nonlinear Operations and Applications, 159–176.

A more thorough examination of the derivation of equation (1) reveals that (1) is actually an approximation of the equation

$$\epsilon\frac{\partial^2 V}{\partial t^2}+(\epsilon f(m,\,h)+1)\frac{\partial V}{\partial t}-\frac{\partial^2 V}{\partial x^2}$$

$$(1')$$

$$=g_{Na}m^3h\,(V_{Na}-V)+g_Kn^4\,(V_K-V)+g_L\,(V_L-V)$$

where ϵ is a positive constant which represents inductance, which is assumed small, and thus the terms involving ϵ are usually ignored. Some numerical studies done by Lieberstein [8] suggest that solutions of (1′), (2)-(4) (with appropriate initial and boundary conditions) do converge, as $\epsilon \to 0$, to solutions of (1), (2)-(4). Thus, some justification is provided for ignoring inductance terms in the usual Hodgkin-Huxley system.

The mathematical justification of these numerical studies is provided in a recent paper of Fitzgibbon and Parrott [5]. Instead of (1′), (2)-(4), they consider the following simplified system with Neumann boundary conditions and the given initial data:

$$\epsilon\frac{\partial^2 u}{\partial t^2}+(\epsilon f_3(w)+1)\frac{\partial u}{\partial t}-\frac{\partial^2 u}{\partial x^2}=f_1(w)\,u+f_2(w)\,,\,t\geq 0\,,\,x\in(0,\,1) \qquad (\mathrm{HH})_\epsilon$$

$$\frac{\partial w}{\partial t}=-kw+g(u,\,w)\,,\,t\geq 0,\,x\in(0,\,1)$$

$$0=\frac{\partial u}{\partial x}(0,\,t)=\frac{\partial u}{\partial x}(1,\,t)\,,\,t\geq 0$$

$$u(x,\,0)=u_0(x)\,,\,x\in(0,\,1)$$

$$\frac{\partial u}{\partial t}(x,\,0)=u_1(x)\,,\,x\in(0,\,1)$$

$$w(x,\,0)=w_0(x)\,,\,x\in(0,\,1)$$

$$\begin{aligned}
&u_0,\, u_1,\, w_0 \in C^\infty(0,\,1)\\
&0 = u_0'(0) = u_0'(1) = u_0'''(0) = u_0'''(1)\\
&0 = u_1'(0) = u_1'(1) = u_1'''(0) = u_1'''(1)\\
&0 = w_0'(0) = w_0'(1) = w_0'''(0) = w_0'''(1).
\end{aligned}$$

Here k is a positive constant, f_1, f_2, f_3 are continuous, polynomial bounded functions, and g is a continuous and uniformly bounded function. (No restrictions on the signs of these functions are made in [5].)

Corresponding to $(HH)_\epsilon$, the usual Hodgkin-Huxley system takes the following form:

$$\frac{\partial u}{\partial t} - \frac{\partial^2 u}{\partial x^2} = f_1(w)\,u + f_2(w)\,,\ t \geq 0,\ x \in (0,\,1) \qquad \text{(HH)}$$

$$\frac{\partial w}{\partial t} = -kw + g(u,\,w)\,,\ t \geq 0,\ x \in (0,\,1)$$

$$0 = \frac{\partial u}{\partial x}(0,\,t) = \frac{\partial u}{\partial x}(1,\,t)\,,\ t \geq 0$$

$$u(x,\,0) = u_0(x)\,,\ x \in (0,\,1)$$

$$w(x,\,0) = w_0(x)\,,\ x \in (0,\,1)$$

$$\begin{aligned}
&u_0,\, w_0 \in C^\infty(0,\,1)\\
&0 = u_0'(0) = u_0'(1) = u_0'''(0) = u_0'''(1)\\
&0 = w_0'(0) = w_0'(1) = w_0'''(0) = w_0'''(1).
\end{aligned}$$

The following lemma gives the boundary behavior of w:

Lemma 1 (5, Lemma 2.8) . *If $(u(\cdot,\,\cdot),\, w(\cdot,\,\cdot))$ is a classical solution to $(HH)_\epsilon$ or (HH), then*

$$[A0 = \frac{\partial w}{\partial x}(0,\,t) = \frac{\partial w}{\partial x}(1,\,t)\,,\ t \geq 0.$$

Fundamental work on the traditional Hodgkin-Huxley equations has been done by Evans [1], [2], [3] and Evans and Shenk [4], and the existence of a unique classical solution of (HH) on $0 \leq t \leq T$, and any $T > 0$, follows from this work. The existence of a unique classical solution of $(\mathrm{HH})_\epsilon$ for each $\epsilon > 0$ on $0 \leq t < \infty$ is shown by using semigroup theory in [5]. We state below the main result of [5], which establishes the convergence of $(\mathrm{HH})_\epsilon$ to (HH) as $\epsilon \to 0$. This result is obtained by using energy arguments, a priori Sobolev estimates, and recent abstract convergence results of Najman [10]. Here $\|\cdot\|_\infty$ denotes the supremum norm of $C[0, 1]$.

Theorem 1 (5, Theorem 6.12) . *Let $(u(\cdot, \cdot), w(\cdot, \cdot))$ be the solution to (HH) and let $(u_\epsilon(\cdot, \cdot), w_\epsilon(\cdot, \cdot))$ be the family of solutions to $(HH)_\epsilon$. Then*

$$\lim_{\epsilon \to 0^+} \|u_\epsilon(\cdot, t) - u(\cdot, t)\|_\infty = 0$$

and

$$\lim_{\epsilon \to 0^+} \|w_\epsilon(\cdot, t) - w(\cdot, t)\|_\infty = 0,$$

$$t \in [0, T], \text{ for any given } T > 0.$$

In the present work we seek to understand the long-term behavior and global dynamics of the perturbed Hodgkin-Huxley system. Specifically, we will show the existence of global attractors A_ϵ for the perturbed system for each ϵ sufficiently small. Properties of these attractors (for example, the fractal dimension and Hausdorff dimension), and the relationship of A_ϵ to A, the global attractor corresponding to the unperturbed Hodgkin-Huxley system, will be the subjects of future work. We note that the existence of the global attractor A follows from known results of Smoller [11] and Temam [12].

We begin by rewriting slightly the equations in $(\mathrm{HH})_\epsilon$ to reflect more accurately the physical system modelled by equations (1′), (2)-(4). We then give physically realistic assumptions which will be sufficient to guarantee our results.

From now on, consider the singularly perturbed Hodgkin-Huxley equations:

$$\epsilon u_{tt} + (N + \epsilon f_3(w)) u_t - u_{xx} + f_1(w) u - f_2(w) = 0, \qquad \text{(SPHH)}$$

$$(t, x) \in \mathbf{R}^+ \times (0, 1),$$

$$w_t = -h_1(u) w + h_2(u),$$

with the Neumann boundary conditions

$$u_x(t, 0) = u_x(t, 1) = 0, t \geq 0, \qquad \text{(NBC)}$$

and initial conditions

$$u(0, x) = u_0(x), u_t(0, x) = u_1(x), w(0, x) = w_0(x), x \in [0, 1].$$

Assumptions: We assume that

1) $N > 0$ is a fixed constant, $\epsilon > 0$ is a constant which can be arbitrarily small.
2) $f_1(w) \geq a$ with $a > 0$ a small constant.
3) $h_1(u) \geq b$ with $b > 0$ a small constant, and $h_2(u) \geq 0$.
4) f_1, f_2, and f_3: $\mathbf{R} \to \mathbf{R}$ and their derivatives are continuous and polynomial bounded.
5) h_1 and h_2: $R \to R$ are uniformly bounded; h_1' and h_2' are locally bounded.
6) $w^* \geq w_0(x) \geq 0$ for $x \in [0, 1]$, where $w^* > 0$ is a fixed constant.

We emphasize that these assumptions concerning the nonlinearity can be actually verified in the case of the real model system. As a consequence of these assumptions, we know that

$$w(t, x) = \exp\left(-\int_0^t h_1(u(\sigma,x))\, d\sigma\right) w_0(x)$$

$$+\int_0^t h_2(u(s,x))\exp\left(-\int_s^t h_1(u(\sigma,x))\,d\sigma\right)ds$$

and therefore,

$$0 \le w(t,x) \le w^* + \frac{C(h_2)}{b} \text{ (denoted by } C_w\text{), for } (t,x) \in \mathbf{R}^+ \times (0,1).$$

Hence, by the assumptions, there exists a uniform constant $C_f > 0$ such that

$$\sup\left\{|f_i(w(t,x))| : (t,x) \in \mathbf{R}^+ \times (0,1)\right\} \le C_f,\ i = 1, 2, 3.$$

From a minor adjustment to the work in [5], for any

$$U_0 = (u_0, u_1, w_0) \in E = H^1(0,1) \times L^2(0,1) \times L^2(0,1),$$

the strong solution $U(t)(x) = (u(t,x), u_t(t,x), w(t,x))$ of the abstract evolution equation

$$\frac{d}{dt}U(t) = G_\epsilon U + F_\epsilon(U),\ t \ge 0,\ U(0) = U_0, \tag{5}$$

exists uniquely and globally for $t \in [0,\infty)$, where $G_\epsilon : D(G_\epsilon)$ $= [H_N^2(0,1) \cap H^1(0,1)] \times H^1(0,1) \times L^2(0,1) \to E$ is the linear operator and F_ϵ is the nonlinear mapping, similar to (4.2) and (4.3) in [5].

Let us recall the definition of a global attractor for a semiflow $\sigma(t,y)$: $\mathbf{R}^+ \times Y \to Y$ in a Banach space Y. Let the semigroup associated with σ be denoted by $S(t) : Y \to Y$. Namely, $S(t)y = \sigma(t,y)$.

A set $\mathsf{A} \subset Y$ is called a *global attractor* for the semiflow σ if it is compact, functional invariant (i.e. $S(t)\mathsf{A} = \mathsf{A}$, for any $t \ge 0$), and dist $(S(t)B, \mathsf{A}) \to 0$ as $t \to 0$, for any bounded set B in Y.

As another related notion, a set $\mathsf{B} \subset Y$ is called an *absorbing set* for the semiflow σ if a flux of trajectories started from any given bounded set $Z \subset Y$ enter into the set B after a transient time duration which is uniform in Z, and will stay in B forever.

We refer a Basic Theorem on the existence of a global attractor for a semiflow to [12] (p.23, Theorem 1.1). Below is the version of this theorem, which we will apply to the semiflow generated by the nonlinear evolution equation (5) which is derived from the original (SPHH) equations.

Basic Theorem. *Assume that the semigroup $S(t)$ in a Banach space Y satisfies the following condition: $S(t) = S_1(t) + S_2(t)$ for every $t \geq 0$, where $S_1(t)$ and $S_2(t) : Y \to Y$ satisfy (i) for every bounded set $Z \subset Y$,* $\sup\{\|S(t)v\| : v \in Z\} \to 0$ *as $t \to +\infty$, (ii) $S_2(t)$ is uniformly compact for t large, i.e. for every bounded set $Z \subset Y$, there is $t_0 \geq 0$, such that*

$$\bigcup_{t \geq t_0} S_2(t) Z \text{ is precompact in } Y.$$

Then there exists a global attractor A in Y for this semiflow σ, provided that there exists a bounded absorbing set B. Moreover, $\mathsf{A} = \omega(\mathsf{B})$ where $\omega(\ .)$ is the ω-limit set. 2

Remark. We note some facts about our system (SPHH) with the Neumann boundary conditions, which might be expected to present obstacles to showing the existence of absorbing sets. First, the (SPHH) system can be viewed as a partially dissipative system of a semilinear hyperbolic equation coupled with ordinary differential equations which have no diffusion term. While there have been some results showing the existence of attractors for partially dissipative semilinear parabolic systems, cf. [9], there are no such results for hyperbolic-type systems as far as the authors are aware. Even for some significant infinite dimensional dynamical systems which are physically considered as dissipative (e.g. 3-dimensional Navier-Stokes equations), the existence of absorbing sets is unknown. Secondly, the Neumann boundary conditions associated with dissipative wave equations or damped elastic Petrovsky equations often lead to difficulties in bounding the mean value of the solutions in the region if there is no additional assumption, cf. [12, Chapter IV].

Here we conduct *a priori* estimates as follows to show that there exist *absorbing sets* in the space E. Let $|.|$ and $\langle .,. \rangle$ denote the norm and the inner-product in $L^2(0, 1)$. Taking the inner-product of the first equation in (SPHH) with u_t and ρu in $L^2(0, 1)$, we have

$$\begin{aligned}&\tfrac{\epsilon}{2}\tfrac{d}{dt}\left|u_t\right|^2+N\left|u_t\right|^2+\epsilon\int_0^1 f_3(w)\left|u_t(t,x)\right|^2dx+\tfrac{1}{2}\tfrac{d}{dt}\left|u_x\right|^2\\&+\int_0^1 f_1(w)\,u(t,x)\,u_t(t,x)\,dx-\left\langle f_2(w),u_t\right\rangle=0,\end{aligned}\tag{6}$$

and

$$\begin{aligned}&\epsilon\rho\tfrac{d}{dt}\left\langle u_t,u\right\rangle-\epsilon\rho\left|u_t\right|^2+\tfrac{N\rho}{2}\tfrac{d}{dt}\left|u\right|^2+\epsilon\rho\int_0^1 f_3(w)\,u(t,x)\,u_t(t,x)\,dx\\&+\rho\left|u_x\right|^2+\rho\int_0^1 f_1(w)\left|u(t,x)\right|^2dx-\rho\left\langle f_2(w),u\right\rangle=0.\end{aligned}\tag{7}$$

Sum up (6) and (7) to obtain

$$\begin{aligned}&\tfrac{1}{2}\tfrac{d}{dt}\left\{\epsilon\left|u_t\right|^2+\left|u_x\right|^2-2\epsilon\rho\left\langle u_t,u\right\rangle+N\rho\left|u\right|^2\right\}\\&+\Bigg\{(N-\epsilon\rho)\left|u_t\right|^2+\rho\left|u_x\right|^2+\epsilon\int_0^1 f_3(w)\left|u_t(t,x)\right|^2dx\\&\quad+\int_0^1\left[f_1(w)+\epsilon\rho f_3(w)\right]u(t,x)\,u_t(t,x)\,dx\\&+\rho\int_0^1 f_1(w)\left|u(t,x)\right|^2dx-\rho\left\langle f_2(w),u\right\rangle-\left\langle f_2(w),u_t\right\rangle\Bigg\}=0.\end{aligned}\tag{8}$$

Define $\Gamma(t)$ by

$$\Gamma(t)=\epsilon\left|u_t\right|^2+\left|u_x\right|^2-2\epsilon\rho\left\langle u_t,u\right\rangle+N\rho\left|u\right|^2.$$

Then we see that

$$\begin{aligned}&\Bigg\{(N-\epsilon\rho)\left|u_t\right|^2+\rho\left|u_x\right|^2+\epsilon\int_0^1 f_3(w)\left|u_t(t,x)\right|^2dx\\&\quad+\int_0^1\left[f_1(w)+\epsilon\rho f_3(w)\right]u(t,x)\,u_t(t,x)\,dx\\&+\rho\int_0^1 f_1(w)\left|u(t,u)\right|^2dx-\rho\left\langle f_2(w),u\right\rangle-\left\langle f_2(w),u_t\right\rangle\Bigg\}-\kappa\Gamma(t)\end{aligned}\tag{9}$$

$$\geq(N-\epsilon\rho-\epsilon\kappa)\left|u_t\right|^2+(\rho-\kappa)\left|u_x\right|^2+\int_0^1\left[f_1(w)+\epsilon\rho f_3(w)+2\epsilon\rho\kappa\right]u(t,x)\,u_t(t,x)\,dx$$

$$+ (\rho a - N\rho\kappa)|u|^2 - C_f(\rho|u| + |u_t|)$$

$$\geq (N - \epsilon\rho - \epsilon\kappa)|u_t|^2 + (\rho - \kappa)|u_x|^2 + (\rho a - N\rho\kappa)|u|^2$$

$$- (C_f + \epsilon\rho C_f + 2\epsilon\rho\kappa)|u||u_t| - C_f(\rho|u| + |u_t|)$$

$$\geq (N - \epsilon\rho - \epsilon\kappa)|u_t|^2 + (\rho - \kappa)|u_x|^2 + (\rho a - N\rho\kappa)|u|^2 - (N/4)|u_t|^2$$

$$- (C_f + \epsilon\rho C_f + 2\epsilon\rho\kappa)^2 |u|^2 / N - (\rho a/2)|u|^2 - (N/4)|u_t|^2 - (C_f)^2\left(\frac{2}{a} + \frac{1}{N}\right)$$

$$\geq \left(\frac{N}{2} - \epsilon\rho - \epsilon\kappa\right)|u_t|^2 + (\rho - \kappa)|u_x|^2$$

$$+ \left(\rho a/2 - N\rho\kappa - (C_f + \epsilon\rho C_f + 2\epsilon\rho\kappa)^2/N\right)|u|^2 - (C_f)^2\left(\frac{2}{a} + \frac{1}{N}\right).$$

Now we have two undetermined constants $\rho > 0$ and $\kappa > 0$ that can be arbitrarily chosen. Besides the intrinsic perturbation parameter $\epsilon > 0$ is a small fixed number. Our idea in making the subsequent ordered choice is to show the following absorbing result.

Theorem 2 *There is a constant $\epsilon_0 > 0$, such that for the semiflow v_ϵ generated by (SPHH) or the formulated abstract evolution equation (5) with parameter $0 < \epsilon \leq \epsilon_0$, there exists an absorbing ball B_R in the space E.*

Proof. In the last inequality of (9), let us first choose and fix $\rho > 0$ such that

$$\frac{\rho a}{2} - N - \frac{4(C_f + 1)^2}{N} \geq 0. \tag{10}$$

Then choose and fix $\kappa > 0$ such that

$$0 < \kappa \leq \min\left\{\rho, \frac{1}{\rho}\right\}, \tag{11}$$

which implies that $\rho - \kappa \geq 0$, and $0 < \rho\kappa \leq 1$. Thus, from (9) it follows that

$$\left\{(N - \epsilon\rho)|u_t|^2 + \rho|u_x|^2 + \epsilon\int_0^1 f_3(w)|u_t(t,x)|^2 dx \right. \tag{12}$$

$$+ \int_0^1 [f_1(w) + \epsilon\rho f_3(w)]\, u(t,x)\, u_t(t,x)\, dx$$

$$\left. +\rho\int_0^1 f_1(w)|u(t,x)|^2 dx - \rho\langle f_2(w), u\rangle - \langle f_2(w), u_t\rangle \right\} - \kappa\Gamma(t)$$

$$\geq \left(\frac{N}{2} - \epsilon\rho - \frac{\epsilon}{\rho}\right)|u_t|^2 + \left(\rho a/2 - N - \frac{(C_f + \epsilon\rho C_f + 2\epsilon)^2}{N}\right)|u|^2 - (C_f)^2\left(\frac{2}{a} + \frac{1}{N}\right)$$

$$\geq -(C_f)^2\left(\frac{2}{a} + \frac{1}{N}\right),$$

where $0 < \epsilon \leq \epsilon_0$, and $\epsilon_0 > 0$ is a constant such that

$$\epsilon_0\left(\rho + \frac{1}{\rho}\right) \leq \frac{N}{2} \quad \text{and} \quad \epsilon_0 \max\{\rho, 1\} \leq 1. \tag{13}$$

Substituting (12) into (8), we get

$$\frac{d}{dt}\Gamma(t) + 2\kappa\Gamma(t) \leq 2(C_f)^2\left(\frac{2}{a} + \frac{1}{N}\right), \tag{14}$$

where $\kappa > 0$ is any constant satisfying (11) in which $\rho > 0$ satisfies (10), and the right-hand side is a constant denoted by

$$C_1 = 2(C_f)^2\left(\frac{2}{a} + \frac{1}{N}\right).$$

Therefore,

$$\Gamma(t) \leq \Gamma(0)\exp(-2\kappa t) + \frac{C_1}{2\kappa}, \text{ for } t \geq 0. \tag{15}$$

The regularity of the initial data allows pointwise evaluation and it is easy to see that $\Gamma(0)$ is a functional of (u_0, u_1) and

$$\Gamma(0) \leq \max\left\{\epsilon_0, 1, N\rho + \rho^2\right\} \|(u_0, u_1)\|^2_{H^1 \times L^2}. \tag{16}$$

On the other hand we have

$$\Gamma(t) \geq \frac{\epsilon}{2}|u_t|^2 + |u_x|^2 + \left(N\rho - 2\epsilon\rho^2\right)|u|^2 \tag{17}$$

where, by (13),

$$N\rho - 2\epsilon\rho^2 = 2\rho\left(\frac{N}{2} - \epsilon\rho\right) \geq 2\epsilon.$$

Therefore,

$$\Gamma(t) \geq \frac{\epsilon}{2}|u_t|^2 + |u_x|^2 + 2\epsilon|u|^2 \geq \frac{\epsilon}{2}\|(u(t), u_t(t))\|^2_{H^1 \times L^2}. \tag{18}$$

Combining (15) with (16) and (18), we obtain

$$\|(u(t), u_t(t))\|^2_{H^1 \times L^2} \leq \frac{2}{\epsilon}\max\left\{\epsilon_0, 1, N\rho + \rho^2\right\} \|(u_0, u_1)\|^2_{H^1 \times L^2}\exp(-2\kappa t) \tag{19}$$

$$+\frac{C_1}{\epsilon\kappa}, \text{ for } t \geq 0.$$

This means that

$$\limsup_{t\to\infty} \|(u(t), u_t(t))\|^2_{H^1 \times L^2} \leq \frac{C_1}{\epsilon\kappa}. \tag{20}$$

Besides it is known that $\|w(t, .)\|^2_{L^2} \leq (C_w)^2$. Incorporating this with (20), finally we can assert that the ball $B_R \subset E$ centered at the origin and with radius R,

$$R = \sqrt{\frac{C_1}{\epsilon\kappa} + (C_w)^2}, \tag{21}$$

is an absorbing set of the semiflow v_ϵ with $0 < \epsilon \le \epsilon_0$. The proof is completed. 2

Theorem 3 . *There exists a global attractor $M \subset E$ for the semiflow v_ϵ generated by (SPHH) or the formulated abstract evolution equation (5) with parameter $0 < \epsilon \le \epsilon_0$, where $\epsilon_0 > 0$ is a uniform constant.*

Proof. Since we have shown the existence of a bounded absorbing set by Theorem 2, it suffices to show that the solution semigroup $S_\epsilon(t)$ associated with the semiflow v_ϵ satisfies the decomposition property required in the Basic Theorem.

For this purpose, we write the solution $w(t)$ of the second equation of (SPHH) as $w(t) = w_1(t, .) + w_2(t, .)$ with

$$\begin{aligned} w_1(t, x) &= w_0(x) \exp\left(-\int_0^t h_1(u(s, x))\, ds\right), \\ w_2(t, x) &= \int_0^t h_2(u(s, x)) \exp\left(-\int_s^t h_1(u(\sigma, x))\, d\sigma\right) ds, \end{aligned} \tag{22}$$

and write the solution $u(t)$ of the first equation of (SPHH) as $u(t) = v_1(t, .) + v_2(t, .)$ with $v_1(t, .)$ satisfying the following linear equation:

$$\begin{gathered} \epsilon v_{tt} + (N + \epsilon f_3(w))\, v_t - v_{xx} + av = 0, \text{ with (NBC)}, \\ v_1(0) = u_0,\ v_{1t}(0) = u_1. \end{gathered} \tag{23}$$

Definite component families of operators $S_{\epsilon 1}(t)$ and $S_{\epsilon 2}(t)$ as a decomposition of $S_\epsilon(t)$ by

$$\begin{aligned} S_{\epsilon 1}(t) &: (u_0, u_1, w_0) \to \left(v_1(t, .), \tfrac{\partial}{\partial t} v_1(t, .), w_1(t, .)\right), \\ S_{\epsilon 2}(t) &: (u_0, u_1, w_0) \to \left(v_2(t, .), \tfrac{\partial}{\partial t} v_2(t, .), w_2(t, .)\right). \end{aligned} \tag{24}$$

It is not difficult to show that $S_{\epsilon 1}$ satisfies the condition (i) in the basic theorem. Here the details are omitted. The remaining task is to check that $S_{\epsilon 2}$ satisfies the uniform compactness for t large, the condition (ii) in the basic theorem.

The component v_2 satisfies the following equation: (we denote v_2 by v for simplicity)

$$\epsilon v_{tt} + N v_t + \epsilon f_3(w) v_t - v_{xx} + f_1(w) u - a v_1 - f_2(w) = 0, \text{ with (NBC)},$$
$$v(0) = 0,\ v_t(0) = 0, \tag{25}$$

where v_1 is the linear component and u is the original solution of the first equation. From the relation $v(t)\,(= v_2(t)) = u(t) - v_1(t)$ and the properties possessed by u and v_1, there is a uniform bound $C_2(r) > 0$ such that

$$\|(v(t),\, v_t(t))\|^2_{H^1 \times L^2} \le C_2(r),\, t \ge 0,$$

for any $(u_0,\, u_1,\, w_0) \in B_r \subset E$.

Taking the inner-product of (25) with $-v_{xxt}$ in $L^2(0,\,1)$, we get

$$\frac{\epsilon}{2}\frac{d}{dt}|v_{xt}|^2 + N|v_{xt}|^2 + \frac{1}{2}\frac{d}{dt}|v_{xx}|^2 \tag{26}$$

$$-\epsilon\frac{d}{dt}\int_0^1 f_3(w) v_t v_{xx} dx + \epsilon\int_0^1 f_3(w) v_{tt} v_{xx} dx + \epsilon\int_0^1 f_3'(w) w_t v_t v_{xx} dx + \frac{d}{dt}\int_0^1 a v_1 v_{xx} dx$$

$$-\frac{d}{dt}\int_0^1 f_1(w) u v_{xx} dx + \frac{d}{dt}\int_0^1 f_2(w) v_{xx} dx - a\int_0^1 v_{1t} v_{xx} dx$$

$$+\int_0^1 f_1(w) u_t v_{xx} dx + \int f_1'(w) w_t u v_{xx} dx - \int_0^1 f_2'(w) w_t v_{xx} dx$$

$$= \frac{d}{dt}\left\{(\epsilon/2)|v_{xt}|^2 + (1/2)|v_{xx}|^2 - \epsilon\int_0^1 f_3(w) v_t v_{xx} dx + \int_0^1 a v_1 v_{xx} dx - \int_0^1 f_1(w) u v_{xx} dx + \int_0^1 f_2(w) v_{xx} dx\right\}$$

$$+N|v_{xt}|^2 + \epsilon\int_0^1 f_3(w) v_{tt} v_{xx} dx + \epsilon\int_0^1 f_3'(w) w_t v_t v_{xx} dx - a\int_0^1 v_{1t} v_{xx} dx$$

$$+\int_0^1 f_1(w) u_t v_{xx} dx + \int_0^1 f_1'(w) w_t u v_{xx} dx - \int_0^1 f_2'(w) w_t v_{xx} dx$$

$$= \frac{d}{dt}\left\{(\epsilon/2)\left|v_{xt}\right|^2 + (1/2)\left|v_{xx}\right|^2 - \epsilon\int_0^1 f_3(w)\, v_t v_{xx} dx + \int_0^1 a v_1 v_{xx} dx - \int_0^1 f_1(w)\, u v_{xx} dx + \int_0^1 f_2(w)\, v_{xx} dx\right\}$$

$$+N\left|v_{xt}\right|^2 - \int_0^1 f_3(w)\left[N v_t - v_{xx} + \epsilon f_3(w)\, v_t + f_1(w)\, u - a v_1 - f_2(w)\right] v_{xx} dx$$

$$+\epsilon\int_0^1 f_3'(w)\, w_t v_t v_{xx} dx - a\int_0^1 v_{1t} v_{xx} dx + \int_0^1 f_1(w)\, u_t v_{xx} dx + \int_0^1 f_1'(w)\, w_t u v_{xx} dx - \int_0^1 f_2'(w)\, w_t v_{xx} dx = 0.$$

Then taking the inner-product of the equation (25) with $-\eta v_{xx}$ in $L^2(0, 1)$, we get

$$\begin{aligned}\langle \epsilon v_{tt} + N v_t + \epsilon f_3(w)\, v_t - v_{xx} + f_1(w)\, u - a v_1 - f_2(w),\ -\eta v_{xx}\rangle \\ = \tfrac{d}{dt}\epsilon\eta\langle v_{xt},\ v_x\rangle - \epsilon\eta\left|v_{xt}\right|^2 \\ +\eta\left|v_{xx}\right|^2 \langle (N + \epsilon f_3(w))\, v_t + f_1(w)\, u - a v_1 - f_2(w),\ -\eta v_{xx}\rangle.\end{aligned} \tag{27}$$

Summing up (26) and (27), we obtain

$$\frac{d}{dt}\Pi(t) + (N - \epsilon\eta)\left|v_{xt}\right|^2 + \eta\left|v_{xx}\right|^2 \le \left|K(t;\ u[B,\ v_1,\ w)\right|\left|v_{xx}\right|, \tag{28}$$

where

$$\Pi(t) = (\epsilon/2)\left|v_{xt}\right|^2 + (1/2)\left|v_{xx}\right|^2 + \epsilon\eta\langle v_{xt},\ v_x\rangle$$

$$-\epsilon\int_0^1 f_3(w)\, v_t v_{xx} dx + \int_0^1 a v_1 v_{xx} dx - \int_0^1 f_1(w)\, u v_{xx} dx + \int_0^1 f_2(w)\, v_{xx} dx,$$

and

$$K(t;\ u,\ v_1,\ w) = f_3(w)\left[N v_t + \epsilon f_3(w)\, v_t + f_1(w)\, u - a v_1 - f_2(w)\right]$$

$$+\epsilon f_3'(w)\,w_t v_t - a v_{1t} + f_1(w)\,u_t + f_1'(w)\,w_t u + f_2'(w)\,w_t$$

$$-\eta\left[(N+\epsilon f_3(w))\,v_t + f_1(w)\,u - a v_1 - f_2(w)\right].$$

We have (19) and a corresponding decay estimate for the component v_1. Besides,

$$|w_t(t,x)| \le |h_1(u(t,x))|\,|w(t,x)| + |h_2(u(t,x))| \le C_h(1+C_w),\ \forall t\ge 0,\ x\in[0,1],$$

where $C_h > 0$ is the uniform bound for both h_1 and h_2. Hence, there is a constant $C_K(r) > 0$, which depends on $r > 0$, such that $|K(t;\,u,\,v_1,\,w)| \le C_K$, for any $(u_0,\,u_1,\,w_0) \in B_r$, with B_r a bounded ball in E centered at the origin and with radius r. Therefore,

$$\frac{d}{dt}\Pi(t) + \frac{N}{2}|v_{xt}|^2 + \eta\,|v_{xx}|^2 \le C_K\,|v_{xx}| \le \frac{\eta}{4}|v_{xx}|^2 + \frac{C_K(r)^2}{\eta}, \tag{29}$$

if $\eta > 0$ is chosen to satisfy

$$0 < \eta \le \frac{N}{2\epsilon_0}. \tag{30}$$

Moreover, for small δ such that $0 < \delta \le \min\{N/\epsilon_0,\ \eta/2\}$, we have

$$\begin{gathered}\tfrac{1}{2}\left[N\,|v_{xt}|^2 + \eta\,|v_{xx}|^2\right] - \delta\Pi(t) = \tfrac{1}{2}(N-\epsilon\delta)\,|v_{xt}|^2 + \tfrac{1}{2}(\eta-\delta)\,|v_{xx}|^2 - \delta\epsilon\eta\,\langle v_{xt},\,v_x\rangle \\ +\delta\epsilon\int_0^1 f_3(w)\,v_t v_{xx}dx - \delta\int_0^1 a v_1 v_{xx}dx + \delta\int_0^1 f_1(w)\,u v_{xx}dx - \delta\int_0^1 f_2(w)\,v_{xx}dx \\ \ge -\delta\,|L(t;\,u,\,v_1,\,w)|\,|v_{xx}|,\end{gathered} \tag{31}$$

where $L(t;\,u,\,v_1,\,w) = \epsilon f_3(w)\,v_t - a v_1 + f_1(w)\,u - f_2(w) - \delta\epsilon\eta v_t$. Similarly, there is a constant $C_L(r) > 0$, which depends on $r > 0$, such that $|L(t;\,u,\,v_1,\,w)| \le C_L(r)$, for any $(u_0,\,u_1,\,w_0) \in B_r$, with B_r a bounded ball in E centered at the origin and with radius r. From (29) and (31) it follows that

$$\frac{d}{dt}\Pi(t) + \delta\Pi(t) + \frac{\eta}{4}|v_{xx}|^2 \le \delta C_L(r)|v_{xx}| + \frac{C_K(r)^2}{\eta}$$

$$\le \frac{\eta}{4}|v_{xx}|^2 + \frac{\delta^2 C_L(r)^2}{\eta} + \frac{C_K(r)^2}{\eta},$$

or

$$\frac{d}{dt}\Pi(t)+\delta\Pi(t) \le \frac{1}{\eta}\left[C_K(r)^2 + \delta^2 C_L(r)^2\right], \text{ for } t \ge 0 \text{ and any } (u_0, u_1, w_0) \in B_r. \tag{32}$$

Because $\Pi(0) = 0$, it follows that

$$\Pi(t) \le \frac{1}{\delta\eta}\left[C_K(r)^2 + \delta^2 C_L(r)^2\right], t \ge 0.$$

Consequently, we have

$$\begin{aligned}(\epsilon/2)|v_{xt}|^2 + (1/2)|v_{xx}|^2 \le \Bigg| & \epsilon\int_0^1 f_3(w)\,v_t v_{xx}dx - \int_0^1 a v_1 v_{xx}dx \\ & + \int_0^1 f_1(w)\,u v_{xx}dx - \int_0^1 f_2(w)\,v_{xx}dx\Bigg| \\ & -\epsilon\eta\langle v_{xt}, v_x\rangle + \tfrac{1}{\delta\eta}\left[C_K(r)^2 + \delta^2 C_L(r)^2\right],\end{aligned} \tag{33}$$

which implies, by the same technique of using Young's inequality to handle the integrals in (33), that

$$|v_t|^2 + |v_{xt}|^2 + |v_{xx}|^2 + |v_x|^2 + |v|^2 = \|(v_2(t), v_2(t))\|^2_{H^2\times H^1} \le C^*(r), \tag{34}$$

for any $(u_0, u_1, w_0) \in B_r \subset E$.

Finally, we treat $w_2(t, x)$ which satisfies the following equation:

$$\frac{d}{dt}w_2 = -h_1(u)\,w_2 + h_2(u), \; w_2(0) = 0, \; t \ge 0. \tag{35}$$

By taking derivatives and inner-product with w_{2x}, we get

$$\frac{1}{2}\frac{d}{dt}|w_{2x}|^2 + b|w_{2x}|^2 \le \frac{1}{2}\frac{d}{dt}|w_{2x}|^2 + \int_0^1 h_1(u)|w_{2x}|^2 dx \\ \le |\langle h_1'(u)\,u_x w_2 + h_2'(u)\,u_x,\, w_{2x}\rangle| \le \frac{b}{2}|w_{2x}|^2 + Q(r), \tag{36}$$

where $Q(r)$ is a constant depending on r, such that for any $(u_0, u_1, w_0) \in B_r$, and for $t \ge 0$, $h_1'(u)|u_x||w_2| + |h_2'(u)||u_x| \le Q(r)$. Thus there is a constant $Q^*(r) > 0$, such that

$$\|w_2\|_{H^1}^2 = |w_{2x}|^2 + |w_2|^2 \le Q^*(r), \tag{37}$$

for $t \ge 0$ and any $(u_0, u_1, w_0) \in B_r$. Combining (30) and (33), we conclude that

$$\bigcup_{t\ge 0} S_{\epsilon 2}(t)\,B_r \subset \text{ a bounded ball in } H^2(0,1) \times H^1(0,1) \times H^1(0,1), \\ \text{centered at the origin and with radius } [C^*(r) + Q^*(r)]^{1/2}. \tag{38}$$

By the Sobolev imbedding theorem, $H^2(0,1) \times H^1(0,1) \times H^1(0,1)$ is embedded in E compactly. Therefore, the component family of operators $S_{\epsilon 2}(t)$ satisfies the condition (ii) in the basic theorem. This completes the proof of Theorem 3. 2

References

[1] J.W. Evans, "Nerve axon equations I: linear approximations," *Indiana Univ. Math. J.*, 21 (1972), 877-885.

[2] J.W. Evans, "Nerve axon equations II: stability at rest," *Indiana Univ. Math. J.*, 22 (1972), 75-90.

[3] J.W. Evans, "Nerve axon equations III: stability of the nerve impulse," *Indiana Univ. Math. J.*, 22 (1972), 577-593.

[4] J.W. Evans and N.A. Shenk, "Solutions to axon equations," *Biophys. J.*, 10 (1970), 1090-1101.

[5] W.E. Fitzgibbon and M.E. Parrott, "Convergence of singularly perturbed Hodgkin-Huxley systems," *Nonlinear Anal.*, TMA, to appear.

[6] J.K. Hale, *Asymptotic Behavior of Dissipative Systems, Amer. Math. Soc.*, Providence, R.I., 1988.

[7] O.A. Ladyzhenskaya, *Attractors for Semigroups and Evolution Equations, Cambridge Univ. Press*, Cambridge, England, 1991.

[8] H.M. Lieberstein, "On the Hodgkin-Huxley partial differential equation," *Math. Biosci.*, 1 (1967), 45-69.

[9] M. Marion, "Finite dimensional attractors to partly dissipative reaction-diffusion systems," *SIAM J. Math. Anal.*, 20 (1989), 816-844.

[10] B. Najman, "Time singular limit of semilinear wave equations with damping," preprint 1990.

[11] J. Smoller, *Shock Waves and Reaction-Diffusion Equations, Springer-Verlag*, New York, 1983.

[12] R. Temam, *Infinite-Dimensional Dynamical Systems in Mechanics and Physics, Springer-Verlag*, New York, 1988.

W. Fitzgibbon
Department of Mathematics
University of Houston
Houston, Texas 77004

M. Parrott
Department of Mathematics
University of South Florida
Tampa, Florida 33620-5700

Y. You
Department of Mathematics
University of South Florida
Tampa, Florida 33620-5700

On strongly elliptic differential operators on $L^1(\mathbb{R}^n)$

Matthias Hieber

1. Introduction

Let $A := \sum_{|\alpha|=m} a_\alpha D^\alpha$ be a homogeneous differential operator with constant coefficients a_α. Then A is called strongly elliptic if

$$Re\, a(\xi) > 0 \quad \text{for} \quad \xi \in \mathbb{R}^n \backslash \{0\}.$$

Here a is the symbol of A defined by $a(\xi) := \sum_{|\alpha|=m} a_\alpha (i\xi)^\alpha$. We remark first that the order m of A is necessarily even, due to the fact that $a(-\xi) = (-1)^m a(\xi)$ for all $\xi \in \mathbb{R}^n$. Second, consider the L^p-realization $\mathcal{A}_p$ of A $(1 \leq p \leq \infty)$ given by

$$(1.1) \qquad \begin{aligned} D(\mathcal{A}_p) &:= \{f \in L^p(\mathbb{R}^n); \mathcal{F}^{-1}(a\hat{f}) \in L^p(\mathbb{R}^n)\} \qquad \text{and} \\ \mathcal{A}_p f &:= \mathcal{F}^{-1}(a\hat{f}) \quad \text{for all} \quad f \in D(\mathcal{A}_p). \end{aligned}$$

Here $\mathcal{F}$ denotes the Fourier transform in S', the space of tempered distributions. Then it is well known that the operator $-\mathcal{A}_p$ generates a holomorphic C_0-semigroup on $L^p(\mathbb{R}^n)$ whenever $1 < p < \infty$. Moreover, this assertion remains valid for the variable coefficient operator $\sum_{|\alpha|\leq m} a_\alpha(x) D^\alpha$, provided that A is uniformly strongly elliptic and $a_\alpha \in L^\infty$ for all $|\alpha| \leq m$ and $a_\alpha \in BUC$ for all $|\alpha| = m$. The crucial argument in the proof of the results mentioned above is Mihlin's multiplier theorem (cf.[S;p.96]), which in particular implies that $D(\mathcal{A}_p) = H_p^m(\mathbb{R}^n)$. Note that in the case $p = 1$ neither Mihlin's theorem nor the above characterization of $D(\mathcal{A}_p)$ remain valid.

G. R. Goldstein and J. A. Goldstein (eds.),
Semigroups of Linear and Nonlinear Operations and Applications, 177–183.

Assuming the Hölder continuity of the coefficients, Guidetti [Gu2] recently proved that operators of this kind generate holomorphic semigroups also on $L^1(\mathbb{R}^n)$. For similar results in this direction, we refer to [Gu1] and [R].

The aim of this paper is to show that if the coefficients of the principal part are constant, i.e. if $a_\alpha \in \mathbb{C}$ for $|\alpha| = m$, then one can give an easy proof of the result mentioned above.

2. Estimates on the principal part

Let A be strongly elliptic. In order to prove that $-\mathcal{A}_1$ generates a holomorphic semigroup on $L^1(\mathbb{R}^n)$, we make use of the following simple sufficient criterion for a function to belong to $\mathcal{M}_1$. Here $\mathcal{M}_1$ denotes the Banach space defined by

$$\mathcal{M}_1 := \{a \in L^\infty(\mathbb{R}^n); \|a\|_{\mathcal{M}_1} < \infty\},$$

where

$$\|a\|_{\mathcal{M}_1} := \sup\{\|\mathcal{F}^{-1}(a\hat{f})\|_1; f \in S(\mathbb{R}^n), \|f\|_1 \leq 1\}.$$

For $\varepsilon > 0$, we denote by $\mathcal{M}^\varepsilon$ the Banach space

$$\mathcal{M}^\varepsilon := \{a \in C^j(\mathbb{R}^n, \mathbb{C}); \|a\|_{\mathcal{M}^\varepsilon} < \infty\},$$

where $j = \min\{k \in \mathbb{N}, k > \frac{n}{2}\}$ and

$$\|a\|_{\mathcal{M}^\varepsilon} := \sup_{|\alpha| \leq j} \sup_{\xi \in \mathbb{R}^n} |\xi|^{\alpha+\varepsilon} |D^\alpha a(\xi)|.$$

Lemma 2.1 *([Hi;Lemma 3.2]. Let $\varepsilon > 0$. Then $\mathcal{M}^\varepsilon \hookrightarrow \mathcal{F}L^1(\mathbb{R}^n) \hookrightarrow \mathcal{M}_1$.*

We remark that if $a \in M^{\varepsilon}$, then the mapping $T_a : f \mapsto \mathcal{F}^{-1}(a\hat{f})$ is a bounded, linear operator on $L^1(\mathbb{R}^n)$ with $\|T_a\| \leq C\|a\|_{\mathcal{M}^{\varepsilon}}$ for some $C > 0$.

Lemma 2.1 implies in particular that the spectrum of $-\mathcal{A}_1$ is contained in a sector of the form

$$S_\theta := \{z \in \mathbb{C}; |arg z - \pi| \leq \theta\}$$

for some θ satisfying $0 \leq \theta < \frac{\pi}{2}$. To be more precise, θ can be calculated to be $\theta = \max_{|\xi|=1} arctg \frac{Im a(\xi)}{Re a(\xi)}$. In particular, if $a_\alpha \in \mathbb{R}$, then $S_\theta = \mathbb{R}_- \cup \{0\}$.

Second, Lemma 2.1 is the key tool in the proof of the following result (see also [Gu2] and [R;Ch.V]). In our proof we follow an idea of Arendt [A].

Proposition 2.2. *If A is strongly elliptic, then $-\mathcal{A}_1$ generates a bounded, holomorphic C_0-semigroup on $L^1(\mathbb{R}^n)$.*

Proof. In view of [N;Thm.AII 1.14], we first prove that

$$\sup_{Re\lambda > 0} \|\lambda(\lambda + \mathcal{A}_1)^{-1}\| < \infty.$$

To this end, let $\lambda = e^{i\phi}$ for some $\phi \in (-\frac{\pi}{2}, \frac{\pi}{2})$ and note that $\lambda(\lambda + \mathcal{A}_1)^{-1} = T_{r_\lambda}$, where $r_\lambda : \mathbb{R}^n \to \mathbb{C}$ is defined by $r_\lambda(\xi) = \lambda(\lambda + a(\xi))^{-1}$. We claim first that $r_\lambda \in \mathcal{M}_1$ with norm bound independent of λ. Differentiating r_λ, we obtain

$$|\xi|^{|\alpha|+1}|D^\alpha r_\lambda(\xi)| \leq C|\xi|^{1-m} \qquad (\xi \in \mathbb{R}^n \backslash \{0\}),$$

where the constant C is independent of λ. Hence, Lemma 2.1 implies that

$$\sup_{Re\lambda > 0, |\lambda|=1} \|\lambda(\lambda + \mathcal{A}_1)^{-1}\| < \infty.$$

Finally, let $\lambda \in \mathbb{C}$ such that $Re\lambda > 0$ and note that $r_{\frac{\lambda}{|\lambda|}}(\xi) = \frac{\lambda}{|\lambda|}(\frac{\lambda}{|\lambda|} + a(\xi))^{-1}$. Define s_λ by $s_\lambda(\xi) := r_{\frac{\lambda}{|\lambda|}}(\frac{\xi}{|\lambda|^{1/m}})$. Since $\mathcal{M}_1$ is invariant under homothetic maps of $\mathbb{R}^n$ (cf.[Hö;Thm.1.13]), we conclude that $s_\lambda \in \mathcal{M}_1$ and that $\|s_\lambda\|_{\mathcal{M}_1} = \|r_{\frac{\lambda}{|\lambda|}}\|_{\mathcal{M}_1}$. Since $s_\lambda(\xi) = \lambda(\lambda + a(\xi))^{-1}$ for all $\lambda \in \mathbb{C}$ with $Re\lambda > 0$, the proof is complete. □

Remark. The proof shows that the assertion of the above proposition remains true also for the spaces $C_0(\mathbb{R}^n)$ and $BUC(\mathbb{R}^n)$.

Note that the domain of $\mathcal{A}_1$ is different from the Sobolev space W_1^m or the Bessel-potential space H_1^m. However, $D(\mathcal{A}_1)$ can be sandwiched by suitable function spaces. To be more precise, for $s \in \mathbb{R}$ define $J^s : \mathbb{R}^n \to \mathbb{R}$ by $J^s(\xi) := (1 + |\xi|^2)^{s/2}$. Then, following the standard notation (cf.[BL]), we set for $m \in \mathbb{N} \cup \{0\}$ and $s \in \mathbb{R}$

$$W_1^m := \{f \in S'(\mathbb{R}^n); D^\alpha f \in L^1(\mathbb{R}^n) \quad \text{for all} \quad |\alpha| \le m\}$$
$$H_1^s := \{f \in S'(\mathbb{R}^n); \mathcal{F}^{-1}(J^s \hat{f}) \in L^1(\mathbb{R}^n)\}.$$

Lemma 2.3. *Let $-\mathcal{A}_1$ be a strongly elliptic operator on $L^1(\mathbb{R}^n)$ of order $2m$. If $0 < \delta < 2m$, then*

$$H_1^{2m+\delta} \hookrightarrow D(\mathcal{A}_1) \hookrightarrow H_1^{2m-\delta}.$$

Moreover, if $s > 2m - 1$, then $H_1^s \hookrightarrow W_1^{2m-1}$.

Proof. In order to prove that $D(\mathcal{A}_1) \hookrightarrow H_1^{2m-\delta}$, it suffices to show that the function $\xi \mapsto J^{2m-\delta}(\xi) r_\lambda(\xi)$ belongs to $\mathcal{M}_1$. Here r_λ is defined by $r_\lambda(\xi) := (\lambda + a(\xi))^{-1}$ for some $\lambda \in \mathbb{C}$ with $Re\lambda > 0$. But this follows easily from Lemma 2.1. On the other hand, let $u \in H_1^{2m+\delta}$. Again Lemma 2.1 implies that the function g given by $\xi \mapsto \xi^\alpha J^{-(2m+\delta)}(\xi)$ belongs to $\mathcal{M}_1$ for all α with $|\alpha| = 2m$. Hence $\mathcal{F}^{-1}(g\hat{u}) \in L^1$ and therefore $u \in D(\mathcal{A}_1)$. The last assertion is similarly easy to prove. □

Remark. We are aware of the stronger result

$$B_{11}^{2m} \subset D(\mathcal{A}_1) \subset B_{1\infty}^{2m},$$

due to Guidetti [Gu1]. Here B_{pq}^s denotes the Besov space defined as in [BL] or [T]. However, the easy accessible assertion of Lemma 2.3 is sufficient for our purposes.

3. The main result

Consider now a differential operator with variable coefficients of the form

$$A = \sum_{|\alpha|=2m} a_\alpha D^\alpha + \sum_{|\alpha|\leq 2m-1} b_\alpha D^\alpha,$$

where $a_\alpha \in \mathbb{C}$ and $b_\alpha \in L^\infty$. We assume the principal part $A^H := \sum_{|\alpha|=2m} a_\alpha D^\alpha$ of A to be strongly elliptic. Moreover, setting $a^H : \mathbb{R}^n \to \mathbb{C}, a^H(\xi) := \sum_{|\alpha|=2m} a_\alpha (i\xi)^\alpha$, we define the L^1-realization of A by

$$D(\mathcal{A}_1) := \{f \in L^1(\mathbb{R}^n); \mathcal{F}^{-1}(a^H \widehat{f}) \in L^1(\mathbb{R}^n)\} \qquad \text{and}$$
$$\mathcal{A}_1 f := Af \quad \text{for all} \quad f \in D(\mathcal{A}_1).$$

Theorem 3.1. *Let A^H be a strongly elliptic operator of order $2m$ and let B be given by $B := \sum_{|\alpha|\leq 2m-1} b_\alpha D^\alpha$, where $b_\alpha \in L^\infty$. Then $-\mathcal{A}_1$ generates a holomorphic C_0-semigroup on $L^1(\mathbb{R}^n)$.*

Proof. Notice first that by Proposition 2.2, the operator $-\mathcal{A}_1^H$ is the generator of a holomorphic semigroup $(T^H(t))_{t\geq 0}$ on $L^1(\mathbb{R}^n)$. Obviously, $\|Bu\|_1 \leq const.\|u\|_{W_1^{2m-1}}$. Let $\delta, \theta \in (0,1)$. Then, by real interpolation (cf. [BL;Thm.6.2.4]), we obtain

$$H_1^{2m-\delta} \hookrightarrow (L^1, H_1^{2m-\delta})_{\theta,1} = B_{11}^{\theta(2m-\delta)} \hookrightarrow H_1^{\theta(2m-\delta)},$$

where $B_{11}^{\theta(2m-\delta)}$ denotes the Besov space defined as in [BL]. Moreover, for every $\varepsilon > 0$, there exists a $C_\varepsilon > 0$ such that

$$\|u\|_{B_{1,1}^{\theta(2m-\delta)}} \leq \varepsilon \|u\|_{H_1^{2m-\delta}} + C_\varepsilon \|u\|_1 \quad (u \in H_1^{2m-\delta}).$$

Furthermore, by Lemma 2.3, we conclude that $H_1^{\theta(2m-\delta)} \hookrightarrow W_1^{2m-1}$ provided
$\theta > \frac{2m-1}{2m-\delta}$. Therefore, by choosing ε small enough, it follows from Lemma 2.3 that

$$\|Bu\|_1 \leq \varepsilon' \|\mathcal{A}_1^H u\|_1 + C_\varepsilon' \|u\|_1$$

for all $u \in H_1^{2m-\delta}$, all $\varepsilon' > 0$ and suitable C_ε'. Hence B is relatively $\mathcal{A}_1^H$-bounded and the claim follows from the classical perturbation theorem for holomorphic semigroups (cf.[Go;Thm.I.6.6]). □

References

[A] W. Arendt, *Linear Evolution Equations.* Lecture Notes, Univerity of Zürich, (1991).

[BL] J. Bergh, J. Löfström, *Interpolation Spaces.* Springer-Verlag Berlin, Heidelberg, New York, (1976).

[Go] J.A. Goldstein, *Semigroups of Linear Operators and Applications.* Oxford University Press, (1985).

[Gu1] D. Guidetti, *On interpolation with boundary conditions.* Math. Z. 207 (1991), 439-460.

[Gu2] D. Guidetti, *On elliptic systems in L^1.* preprint 1991.

[Hi] M. Hieber, *Integrated semigroups and differential operators in L^p spaces.* Math. Ann. 291 (1991), 1-16.

[Hö] L. Hörmander, *Estimates for translation invariant operators in L^p spaces.* Acta Math. 104 (1960), 93-140.

[N] R. Nagel (ed.), *One-parameter Semigroups of Positive Operators.* Lecture Notes in Math. 1184, Springer-Verlag, Berlin, Heidelberg, New York, Tokyo, (1986).

[R] D.W. Robinson, *Elliptic operators and Lie groups.* Clarendon Press, Oxford, New York, Tokyo (1991).

[S] E. M. Stein, *Singular Integrals and Differentiability Properties of Functions.* Princeton University Press, New Jersey, 1970.

[T] H. Triebel, *Theory of Function Spaces.* Birkhäuser Verlag, Basel, Boston, Stuttgart (1983).

Mathematisches Institut
Universität Zürich
Rämistrasse 74
CH-8001 Zürich
Switzerland

Stability and local invariant manifolds in fully nonlinear parabolic equations

Alessandra Lunardi [1]

1. Introduction

We study the asymptotic behavior for small initial data of the solutions of a class of evolution equations in general Banach space X:

$$u'(t) = Au(t) + F(u(t)), \quad t \geq 0, \tag{1}$$

$$u(0) = u_0. \tag{2}$$

Here $A : D(A) \subset X \mapsto X$ is a sectorial operator, and $F : D_A(\theta + 1, \infty) \mapsto D_A(\theta, \infty)$ is a nonlinear sufficiently smooth function, such that $F(0) = 0$, $F'(0) = 0$. We recall that $D_A(\theta, \infty)$ is the usual real interpolation space between X and $D(A)$, and $D_A(\theta + 1, \infty) = \{x \in D(A) : Ax \in D_A(\theta, \infty)\}$ is the domain of the part of A in $D_A(\theta, \infty)$.

In particular, we state existence and regularity results for initial data in $D_A(\theta + 1, \infty)$, we prove that the principle of linearized stability holds, and we construct local stable, unstable, and center manifolds.

Such a type of nonlinearity arises in the study of fully nonlinear parabolic equations and systems. As a basic example, we consider

$$\begin{cases} u_t(t,x) = \Delta u + cu + f(u, Du, D^2u), & t \geq 0, \ x \in \overline{\Omega}, \\ \partial u / \partial \nu = 0, & t \geq 0, \ x \in \partial\Omega, \end{cases} \tag{3}$$

where Ω is a bounded open set in $\mathbf{R}^n$ with regular boundary $\partial\Omega$, Δ is the Laplace operator, ν is the exterior normal vector to $\partial\Omega$, and $(u, p, q) \mapsto f(u, p, q)$ is defined and twice continuously differentiable in a neighborhood of 0 in $\mathbf{R} \times \mathbf{R}^n \times \mathbf{R}^{n^2}$, $f(0) = 0$, $Df(0) = 0$.

[1]Partially supported by GNAFA of CNR.

G. R. Goldstein and J. A. Goldstein (eds.),
Semigroups of Linear and Nonlinear Operations and Applications, 185–203.

As usual in fully nonlinear equations, we choose $X = C(\overline{\Omega})$. Then the realization A of $\Delta + cI$ with homogeneous boundary condition in X is sectorial. The domain $D(A)$ is defined by

$$D(A) = \left\{ \varphi \in \bigcap_{p \geq 1} W^{2,p}(\Omega) : \Delta\varphi \in X, \ \partial\varphi/\partial\nu = 0 \text{ in } \partial\Omega \right\},$$

and it consists of twice continuously differentiable functions only if the space dimension n is equal to 1. Therefore, the composition $F(\varphi) = f(\varphi, D\varphi, D^2\varphi)$ is not well defined on $D(A)$ if $n > 1$, but it is well defined on the space

$$D_A(\theta + 1, \infty) = \{\varphi \in C^{2+2\theta}(\overline{\Omega}) : \partial\varphi/\partial\nu = 0\},$$

for every $\theta \in]0, 1/2[$, and it has values in $D_A(\theta, \infty) = C^{2\theta}(\overline{\Omega})$.

We use some results of [7], where local existence and regularity for small initial data $u_0 \in D_A(\theta + 1, \infty)$ were proved. In particular, the solution of (1)-(2) is represented by the variation of constants formula

$$u(t) = e^{tA}u_0 + \int_0^t e^{(t-s)A} F(u(s))ds, \quad t \geq 0, \tag{4}$$

which lets us adapt to our situation the Lyapunov-Perron approach for the construction of invariant manifolds. The difficulties arising from the fully nonlinear character of (1) are overcome by using optimal regularity results (and the corresponding estimates) for the linearized problem

$$v'(t) = Av(t) + f(t),$$

both for forward and for backward solutions.

The construction of the center-unstable manifold, in the general case, relies on the assumption that the elements of the spectrum of A with non-negative real part are a finite number of eigenvalues, with finite algebraic multiplicity. If this assumption fails to be satisfied - for instance, if the set Ω in example (3) is unbounded - other techniques have been used in the study of the stability of the null solution: see [7] and the references quoted therein.

The results about problem (1) are similar to the ones of [3], where the continuous interpolation spaces $D_A(\theta)$, $D_A(\theta+1)$ were used instead of $D_A(\theta, \infty)$, $D_A(\theta + 1, \infty)$. We recall that $D_A(\theta)$ is the closure of $D(A)$ in $D_A(\theta, \infty)$, and

$D_A(\theta+1) = \{x \in D(A) : Ax \in D_A(\theta)\}$ is the domain of the part of A in $D_A(\theta)$. In the applications to parabolic problems such as (3), $D_A(\theta)$ turns out to be the space $h^{2\theta}(\overline{\Omega})$ of the little-Hölder continuous functions of exponent 2θ, which is the closure of $C^\infty(\overline{\Omega})$ in $C^{2\theta}(\overline{\Omega})$. Correspondingly, $D_A(\theta+1) = \{\varphi \in C^2(\overline{\Omega}) : D_{ij}\varphi \in h^{2\theta}(\overline{\Omega}),\ \partial\varphi/\partial\nu = 0\}$.

It seems useful to develop a theory which avoids the little-Hölder spaces and employs only the more familiar Hölder spaces. Moreover, we do not follow closely the proofs of [3] but we simplify and clarify them. In particular, we find new existence in the large results (Proposition 9).

The abstract results are applied to the study of the behavior of the solutions of problems of the type (3) (also with Dirichlet boundary condition) for small initial data. Concerning (3), for $c < 0$ the null solution is exponentially asymptotically stable in $C^{2+2\theta}(\overline{\Omega})$, if $c > 0$ the null solution is unstable, and we prove the existence of an infinite dimensional stable manifold and of a finite dimensional unstable or center-unstable manifold. For $c = 0$, we show that the subset of $C^{2+2\theta}(\overline{\Omega})$ consisting of the functions with zero normal derivative and zero mean value is a local center manifold. In particular, this fact reduces the problem of stability for (3) to the problem of stability for the ordinary differential equation $u' = f(u,0,0)$. Therefore, sufficient conditions for stability are easily found: if

$$\partial^j f/\partial u^j(0,0,0) = 0 \quad j = 0,\ldots,k, \quad \partial^k f/\partial u^j(0,0,0) \neq 0,$$

and either k is even, or k is odd and $\partial^k f/\partial u^k(0,0,0) > 0$, then the null solution is unstable; if k is odd and $\partial^k f/\partial u^k(0,0,0) < 0$, then the null solution is stable in $C^{2+2\theta}(\overline{\Omega})$.

2. The principle of linearized stability, the stable and the unstable manifolds

Let X be a Banach space with norm $\|\cdot\|$, and let $A : D(A) \subset X \mapsto X$ be a linear operator such that there are constants $\omega \in \mathbf{R}$, $\theta \in]\pi/2, \pi[$, $M > 0$ satisfying

$$\rho(A) \supset S = \{\lambda \neq \omega : |\arg(\lambda-\omega)| < \theta\}, \quad \|R(\lambda, A)\|_{L(X)} \leq \frac{M}{|\lambda-\omega|} \quad \forall\lambda \in S. \tag{5}$$

Then A generates an analytic semigroup e^{tA} in X (see [11]). The domain $D(A)$ is endowed with the graph norm $\|x\|_{D(A)} = \|x\| + \|Ax\|$.

The interpolation spaces $D_A(\theta,\infty)$, $D_A(\theta+1,\infty)$ are defined, for $0<\theta<1$, by

$$\begin{cases} D_A(\theta,\infty) = \{x \in X \,:\, |x|_{D_A(\theta,\infty)} = \sup_{0<t\le 1} \|t^{1-\theta}Ae^{tA}x\| < \infty\}, \\ \\ \|x\|_{D_A(\theta,\infty)} = \|x\| + |x|_{D_A(\theta,\infty)}; \end{cases}$$

$$\begin{cases} D_A(\theta+1,\infty) = \{x \in D(A) \,:\, \; Ax \in D_A(\theta,\infty)\}, \\ \\ \|x\|_{D_A(\theta+1,\infty)} = \|x\|_{D(A)} + |Ax|_{D_A(\theta,\infty)}. \end{cases}$$

A local existence and uniqueness theorem for the solution of (1)–(2) has been shown in [7].

Theorem 1 *Let $A : D(A) \subset X \mapsto X$ satisfy (5). Let Ω be a neighborhood of 0 in $D_A(\theta+1,\infty)$, with $0<\theta<1$, and let $F : \Omega \mapsto D_A(\theta,\infty)$ be a continuously differentiable function. Then for every $T>0$ there are $R = R(T)$, $R_0 = R_0(T)$, $M = M(T) > 0$ such that if $\|u_0\|_{D_A(\theta+1,\infty)} \le R$, then problem (1)–(2) has a solution $u \in C(]0,T];D_A(\theta+1,\infty)) \cap C^\theta([0,T];D(A))$ $\cap\, C^1(]0,T];D_A(\theta,\infty))$, such that*

$$\sup_{0\le t\le T} \|u(t)\|_{D_A(\theta+1,\infty)} + \|u\|_{C^\theta([0,T];D(A))} \le M\|u_0\|_{D_A(\theta+1,\infty)}. \tag{6}$$

u is the unique solution of (1)–(2) such that $\sup_{0\le t\le T} \|u(t)\|_{D_A(\theta+1,\infty)} \le R_0$.

As a corollary, a continuation result follows.

Corollary 2 *Set $R = \sup_{T>0} R(T)$, where $R(T)$ is given by Theorem 1. Let u be a solution of (1) and set $\tau = \sup\{t>0 \,:\, u \in C(]0,t];D_A(\theta+1,\infty))$ solves (1) in $]0,t]\}$. If $\|u(t)\|_{D_A(\theta+1,\infty)} < R$ for every $t<\tau$, then $\tau = +\infty$.*

Theorem 1 and its corollary will be used later. The results concerning the principle of linearized stability are independent of them. We begin with the stability theorem, which looks similar to the linearized stability theorem for semilinear ([5]) or quasilinear equations ([4], [6], [10]).

Theorem 3 *Let $A : D(A) \subset X \mapsto X$ be a linear operator satisfying (5), such that*

$$\sup\{\operatorname{Re}\lambda : \lambda \in \sigma(A)\} = -\omega_0 < 0. \tag{7}$$

Then for every $\omega \in]0, \omega_0[$ there are r, M such that if $\|u_0\|_{D_A(\theta+1,\infty)} \le r$, then the solution u of (1)–(2) is defined in $[0, +\infty[$, and

$$\|u(t)\|_{D_A(\theta+1,\infty)} + \|u'(t)\|_{D_A(\theta,\infty)} \le Me^{-\omega t}\|u_0\|_{D_A(\theta+1,\infty)}, \quad t \ge 0.$$

The proof of Theorem 3 relies on a similar result for the linear case. Since we deal with exponentially decaying functions, we introduce some weighted spaces. If Y is any Banach space and $\omega \ge 0$, we set

$$\begin{aligned} C_\omega(]0,+\infty[;Y) &= \{u \in C(]0,+\infty[;Y) : t \mapsto e^{\omega t}u(t) \in L^\infty(0,+\infty;Y)\}, \\ \|u\|_{C_\omega(]0,+\infty[;Y)} &= \sup_{t>0} \|e^{\omega t}u(t)\|_Y. \end{aligned}$$

Proposition 4 *Assume that A satisfies (5) and (7). Let $0 < \theta < 1$. Then for every $u_0 \in D_A(\theta+1,\infty)$ and $f \in C_\omega(]0,+\infty[;D_A(\theta,\infty))$, the solution of problem*

$$v'(t) = Av(t) + f(t), \quad t \ge 0; \quad v(0) = u_0, \tag{8}$$

belongs to $C_\omega(]0,+\infty[;D_A(\theta+1,\infty)) \cap C^1(]0,+\infty[;D_A(\theta,\infty))$, and there is $C = C(\omega) > 0$, not depending on f, u_0, such that

$$\|v\|_{C_\omega(]0,+\infty[;D_A(\theta+1,\infty))} \le C(\|u_0\|_{D_A(\theta+1,\infty)} + \|f\|_{C_\omega(]0,+\infty[;D_A(\theta,\infty))}). \tag{9}$$

The proof of Proposition 4 follows the same arguments of [9], Proposition 3.10, where the nonautonomous periodic case $A = A(t)$ is treated. The details of this (simpler) case will appear in [8].

Proof of Theorem 3 — Once Proposition 4 has been established, the proof of Theorem 3 is quite similar to the proof of local existence and uniqueness given in [7]. Indeed, every regular solution of (1)–(2) in $[0,+\infty[$ is a fixed point of the operator Γ defined by

$$(\Gamma u)(t) = e^{tA}u_0 + \int_0^t e^{(t-s)A}F(u(s))ds, \quad t \geq 0.$$

Using Theorem 4, it is easy to see that if R_0 is sufficiently small, then Γ is a 1/2-contraction on the ball $B(0,R_0) \subset C_\omega(]0,+\infty[; D_A(\theta+1,\infty))$. Setting

$$M_{\theta,\omega} = \sup_{t>0} \|e^{\omega t}e^{tA}\|_{L(D_A(\theta+1,\infty))},$$

we get $\|\Gamma u\|_{C_\omega(]0,+\infty[;D_A(\theta+1,\infty))} \leq M_{\theta,\omega}\|u_0\|_{D_A(\theta+1,\infty)} + R_0/2$, so that Γ maps $B(0,R_0)$ into itself provided $\|u_0\|_{D_A(\theta+1,\infty)} \leq r = (2M_{\theta,\omega})^{-1}R_0$, in which case it has a unique fixed point u in $B(0,R_0)$, which is the solution of (1)–(2) enjoying the claimed properties. ∎

For the instability result, the assumption corresponding to (7) is the following:

$$\sigma_+(A) = \sigma(A) \cap \{\lambda \in \mathbf{C} : \mathrm{Re}\lambda > 0\} \neq \emptyset, \quad \inf\{\mathrm{Re}\,\lambda : \lambda \in \sigma_+(A)\} = \omega_+ > 0. \tag{10}$$

If (10) holds, one can define the projection

$$P = \frac{1}{2\pi i}\int_\gamma (\lambda - A)^{-1}d\lambda,$$

where γ is any regular curve surrounding $\sigma_+(A)$, lying in the half plane { Re $\lambda > 0$}, oriented counterclockwise. Then $P(X) \subset D(A^n)$ for every $n \in \mathbf{N}$. We use again spaces of exponentially decaying functions, defined as follows: if Y is any Banach space and $\omega > 0$ we set

$$\begin{aligned} C_\omega(]-\infty,0];Y) &= \{u \in C(]-\infty,0];Y) : t \mapsto e^{\omega t}u(t) \in L^\infty(-\infty,0;Y)\}, \\ \|u\|_{C_\omega(]-\infty,0];Y)} &= \sup_{t<0} \|e^{\omega t}u(t)\|_Y. \end{aligned}$$

Theorem 5 *Let A, F satisfy the assumptions of Theorem 1 and assume that (10) holds. Then there exist nontrivial backward solutions $u \in C(]-\infty,0];D_A(\theta+1,\infty))$ of (1), going to 0 as t goes to $-\infty$. In particular, the* *null solution is unstable. More precisely, fixed any* $\omega \in]0,\omega_+[$ *there are* r_0, r_1, $M > 0$ *and a Lipschitz continuous function*

$$\varphi : B(0,r_0) \subset P(X) \mapsto B(0,r_1) \subset (I-P)(D_A(\theta+1,\infty)),$$

differentiable at 0 *with* $\varphi'(0) = 0$, *such that for every* u_0 *belonging to the graph of* φ *problem (1) has a unique backward solution* $u \in C_\omega(]-\infty,0]; D_A(\theta+1,\infty))$ *with* $\|u\|_{C_\omega(]-\infty,0];D_A(\theta+1,\infty))} \le M$, *and, conversely, if problem (1) has a backward solution* $u \in C_\omega(]-\infty,0]; D_A(\theta+1,\infty))$ *with norm less than* M, *then* $u(0)$ *belongs to the graph of* φ.

The graph of φ is called *unstable manifold.* As in the stability case, the proof of Theorem 5 relies on a result about backward solutions of linear problems, whose proof is quite similar to Proposition 3.11(ii) of [9], and is omitted.

Proposition 6 *Assume that* A *satisfies (5) and (10). Let* $0 < \theta < 1$, *and fix* $\omega \in]0,\omega_+[$. *Then for every* $f \in C_\omega(]-\infty,0]; D_A(\theta,\infty))$, *all the solutions* $z \in C_\omega(]-\infty,0]; D_A(\theta+1,\infty))$ *of problem*

$$z'(t) = Az(t) + f(t), \;\; t \le 0,$$

are given by the representation formula

$$z(t) = e^{tA}x + \int_0^t e^{(t-s)A}Pf(s)ds + \int_{-\infty}^t e^{(t-s)A}(I-P)f(s)ds, \;\; t \le 0,$$

where x *is any element of* $P(X)$. *Moreover, they are continuously differentiable with values in* $D_A(\theta,\infty)$, *and there is* $C > 0$, *not depending on* f, x, *such that*

$$\|z\|_{C_\omega(]-\infty,0];D_A(\theta+1,\infty))} \le C(\|x\| + \|f\|_{C_\omega(]-\infty,0];D_A(\theta,\infty))}). \tag{11}$$

Proof of Theorem 5 — Fixed $x \in P(X)$, we look for a fixed point of the operator Λ defined by

$$(\Lambda_{x,M}u)(t) = e^{tA}x + \int_0^t e^{(t-s)A}PF(u(s))ds + \int_{-\infty}^t e^{(t-s)A}(I-P)F(u(s))ds,$$

in a small ball $B(0,M) \subset C_\omega(]-\infty,0];D_A(\theta+1,\infty))$, with $0<\omega<\omega_+$. Using estimate (11) it is not difficult to see that $\Lambda_{x,M}$ is a 1/2-contraction in $B(0,M)$ if M is sufficiently small; moreover it maps $B(0,M)$ into itself provided $\|x\| \leq r_0 = M/2C$ (C is the constant in estimate (11)). The first part of the statement follows. Concerning the second part, we set

$$\varphi(x) = (I-P)u(0) = \int_{-\infty}^{0} e^{-sA}(I-P)F(u(s))ds,$$

where u is the fixed point of $\Lambda_{x,M}$. Since the mapping $(x,u)\mapsto \Lambda u$ is Lipschitz continuous in $B(0,r_0)\times B(0,M)\subset P(X)\times C_\omega(]-\infty,0];D_A(\theta+1,\infty))$, by the contraction principle depending on a parameter, also φ is Lipschitz continuous. Moreover, since $\Lambda_{x,M}$ is a 1/2- contraction, it holds

$$\|u\|_{C_\omega(]-\infty,0];D_A(\theta+1,\infty))} \leq 2C\|x\|,$$

so that

$$\|\varphi(x)\| \leq C\sup\{\|F'(y)\|_{L(D_A(\theta+1,\infty),D_A(\theta,\infty))} : \|y\|_{D_A(\theta+1,\infty)} \leq M\}2C\|x\|$$

Letting $M\to 0$, since $F'(0)=0$ we find $\varphi'(0)=0$. ∎

In the saddle point case, that is when the spectrum of A does not intersect the imaginary axis, also a local stable manifold may be easily constructed. We need a linear result similar to the one of Proposition 6, for whose proof we refer to [9, Proposition 3.10] or to [8].

Proposition 7 *Assume that A satisfies (5) and that*

$$\sigma(A)\cap i\mathbf{R} = \emptyset. \tag{12}$$

Let $0<\theta<1$, and fix $\omega\in\,]0,\omega_-[$, where $-\omega_- = \sup\{\operatorname{Re}\lambda : \lambda\in\sigma(A),\ \operatorname{Re}\lambda<0\}$. Then for every $f\in C_\omega(]0,+\infty[;D_A(\theta,\infty))$, all the solutions of problem

$$v'(t) = Av(t)+f(t),\quad t\geq 0,$$

belonging to $C_\omega(]0,+\infty[;D_A(\theta+1,\infty))$, are given by the representation formula

$$v(t) = e^{tA}x + \int_0^t e^{(t-s)A}(I-P)f(s)ds - \int_t^{+\infty} e^{(t-s)A}Pf(s)ds,\quad t\geq 0,$$

where x is any element of $(I-P)(D_A(\theta+1,\infty))$. Moreover, they are continuously differentiable in $]0,+\infty[$ with values in $D_A(\theta,\infty)$, and there is $C=C(\omega)>0$, not depending on f, x, such that

$$\|v\|_{C_\omega(]0,+\infty[;D_A(\theta+1,\infty))} \leq C(\|x\|+\|f\|_{C_\omega(]0,+\infty[;D_A(\theta,\infty))}). \tag{13}$$

Using proposition 7 and arguing as in the proof of Theorem 5, one gets

Theorem 8 *Let A, F satisfy the assumptions of Theorem 1 and in addition assume that (12) holds. Then, fixed any $\omega\in\,]0,\omega_-[$ there are r_0, r_1, $M>0$ and a Lipschitz continuous function*

$$\psi : B(0,r_0)\subset (I-P)(D_A(\theta+1,\infty)) \mapsto B(0,r_1)\subset P(X),$$

differentiable at 0 with $\psi'(0)=0$, such that for every u_0 belonging to the graph of ψ problem (1) has a unique solution $u\in C_\omega(]0,+\infty[;D_A(\theta+1,\infty))$ such that $\|u\|_{C_\omega(]0,+\infty[;D_A(\theta+1,\infty))}\leq M$, and, conversely, if problem (1) has a solution $u\in C_\omega(]0,+\infty[;D_A(\theta+1,\infty))$ such that $\|u\|_{C_\omega(]0,+\infty[;D_A(\theta+1,\infty))}\leq M$, then $u(0)$ belongs to the graph of ψ.

The graph of ψ is called *stable manifold.*

3. The center and center-unstable manifolds

Throughout the section we shall assume that

$$\begin{cases} \textit{the set } \sigma_+(A)=\{\lambda\in\sigma(A):\ \mathrm{Re}\,\lambda\geq 0\} \\ \textit{consists of a finite number of isolated eigenvalues} \\ \textit{with finite algebraic multiplicity.} \end{cases} \tag{14}$$

We set as before $\omega_- = -\sup\{\mathrm{Re}\,\lambda:\ \lambda\in\sigma(A),\ \mathrm{Re}\,\lambda<0\}$, and $P=\frac{1}{2\pi i}\int_\gamma(\lambda-A)^{-1}d\lambda$, where γ is any regular curve surrounding $\sigma_+(A)=\{\lambda\in\sigma(A),\ \mathrm{Re}\,\lambda\geq 0\}$, lying in the half plane $\{\ \mathrm{Re}\,\lambda>-\omega_0\}$, oriented counterclockwise.

Then problem (1) is equivalent to the system

$$\begin{cases} x'(t) = A_+x(t) + PF(x(t)+y(t)), & t \geq 0, \\ y'(t) = A_-y(t) + (I-P)F(x(t)+y(t)), & t \geq 0, \end{cases} \tag{15}$$

with $x(t) = Pu(t)$, $y(t) = (I-P)u(t)$, $A_+ = A_{|P(X)} : P(X) \mapsto P(X)$, $A_- = A_{|(I-P)(D_A(\theta+1,\infty))} : (I-P)(D_A(\theta+1,\infty)) \mapsto (I-P)(D_A(\theta,\infty))$. As usual, we modify F by introducing a smooth cutoff function $\rho : P(X) \mapsto \mathbf{R}$, such that $0 \leq \rho(x) \leq 1$, $\rho(x) = 1$ if $\|x\| \leq 1/2$, $\rho(x) = 0$ if $\|x\| \geq 1$, and for small $\rho > 0$ we consider the system

$$\begin{cases} x'(t) = A_0x(t) + f(x(t),y(t)), & t \geq 0, \\ y'(t) = A_-y(t) + g(x(t),y(t)), & t \geq 0, \end{cases} \tag{16}$$

where

$$f(x,y) = PF(\rho(x/r)x+y), \quad g(x,y) = (I-P)F(\rho(x/r)x+y).$$

System (16) coincides with (15) if $\|x(t)\| \leq r/2$. In particular, they are equivalent as far as stability of the null solution is concerned.

Theorem 1 and Corollary 2 may be applied to problem (16), getting local existence for small initial data, and existence in the large provided the solution remains small enough. In fact, while it is easy to find an *a priori* estimate on $\|y(t)\|_{D_A(\theta+1,\infty)}$ for r small (see Proposition 9 below), in general $x(t)$ is not necessarily bounded (for instance, in the case where $f \equiv 0$ and $\|e^{tA}P\|_{L(X)}$ is not bounded there are arbitrarily small x_0 such that $x(t)$ is not bounded). Therefore, Corollary 2 is of no use here. However, due to the truncation in f and g, we can show that if r and the initial data are small enough, then the solution of (16) exists in the large.

Proposition 9 *There are r_0, $C_0 = C_0(r)$, such that for $r \leq r_0$ and $\|x_0\| + \|y_0\|_{D_A(\theta+1,\infty)} \leq C_0(r)$, problem (16) has a solution $(x,y) \in C([0,+\infty[;P(X)) \times C(]0,+\infty[;D_A(\theta+1,\infty)) \cap L^\infty(0,+\infty;D_A(\theta+1,\infty)) \cap C([0,+\infty[;D(A))$. The solution is unique in the class of the functions (x,y) enjoying the above properties of regularity and such that $\|y(t)\|_{D_A(\theta+1,\infty)} \leq r$.*

Proof — The proof is in two steps. First we show that for any fixed $T > 0$, system (16) has a solution in $[0, T]$ provided r is small and $\|y_0\|_{D_A(\theta+1,\infty)} \leq r/2M$, where

$$M = \sup_{t>0} \|e^{tA_-}\|_{L((I-P)(D_A(\theta+1,\infty)))},$$

no matter how $\|x_0\|$ is large. Second, we prove that $\|y(t)\|_{D_A(\theta+1,\infty)} \leq r/2M$ for every t in the maximal interval of existence, provided $\|x_0\|$ and $\|y_0\|_{D_A(\theta+1,\infty)}$ are small enough.

For every $r > 0$, denote by $L(r)$ the maximum between the Lipschitz constants of f and g over $P(X) \times B(0,r) \subset (I-P)(D_A(\theta+1,\infty))$ (obviously, $P(X)$ is endowed with the norm of X and $(I-P)(D_A(\theta+1,\infty))$ is endowed with the norm of $D_A(\theta+1,\infty)$). Then

$$\lim_{r\to 0} L(r) = 0.$$

Let moreover C_- be the constant of estimate (9), with A replaced by A_- and $\omega = 0$, and let $C_+ = C_+(T)$ be such that

$$\left\| \int_0^t e^{(t-s)A_+} f(s)ds \right\| \leq C_+ \|f\|_{L^\infty(0,T;X)}, \quad 0 \leq t \leq T,$$

for every $f \in L^\infty(0,T;P(X))$.

Any solution of (16) is a fixed point of the operator Γ defined by

$$\begin{aligned}\Gamma(x,y)(t) &= \Big(e^{tA_+}x_0 + \int_0^t e^{(t-s)A_+} f(x(s),y(s))ds, \\ &\qquad e^{tA_-}y_0 + \int_0^t e^{(t-s)A_-} g(x(s),y(s))ds\Big).\end{aligned}$$

If r is small enough, Γ is well defined on the set $Y = B(e^{tA_+}x_0, r) \times B(0,r) \subset C([0,T];P(X)) \times C(]0,T];(I-P)(D_A(\theta+1,\infty)) \cap L^\infty(0,T;(D_A(\theta+1,\infty))$. Y is endowed with the product norm $\|(x,y)\|_Y = \|x\|_{L^\infty(0,T;X)} + \|y\|_{L^\infty(0,T;D_A(\theta+1,\infty))}$.

For (x,y), $(\overline{x},\overline{y})$ in Y we have $\|f(x(s),y(s)) - f(\overline{x}(s),\overline{y}(s))\| \leq L_r\ (\|x(s) - \overline{x}(s)\| + \|y(s) - \overline{y}(s)\|_{D_A(\theta+1,\infty)})$, $\|g(x(s),y(s)) - g(\overline{x}(s),\overline{y}(s))\|_{D_A(\theta,\infty)} \leq L_r\ (\|x(s) - \overline{x}(s)\| + \|y(s) - \overline{y}(s)\|_{D_A(\theta+1,\infty)})$. Therefore, Γ is a 1/2-contraction provided

$$(C_+ + C_-)L_r \leq 1/2.$$

Moreover, setting $M(r) = \sup_{x\in P(X),\|y\|_{D_A(\theta+1,\infty)}\le r} \|f(x,y)\| + \|g(x,y)\|_{D_A(\theta,\infty)}$, we have

$$\lim_{r\to 0} M(r)/r = 0.$$

For $(x, y) \in Y$ we get

$$\|\Gamma(x,y) - (e^{tA_+}x_0, 0)\|_Y \le (C_+ + C_-)M(r) + M\|y_0\|_{D_A(\theta+1,\infty)}.$$

In particular, Γ maps Y into itself if r is so small that $(C_+ + C_-)M(r)/r \le 1/2$ and $\|y_0\|_{D_A(\theta+1,\infty)} \le r/2M$. So, if r is small enough and $\|y_0\|_{D_A(\theta+1,\infty)} \le r/2M$, problem (16) has a unique solution in Y. Indeed, it is the unique solution such that $\|y(t)\|_{D_A(\theta+1,\infty)} \le r$.

Let us prove now that $\|y(t)\|_{D_A(\theta+1,\infty)} \le r/2M$ for every t in the interval of existence if the initial data are small enough. By Theorem 1, there is $C(r)$ such that if $\|x_0\| + \|y_0\|_{D_A(\theta+1,\infty)} \le C(r)$, then the solution of (16) is defined at least in $[0,1]$, and $\|y(t)\|_{D_A(\theta+1,\infty)} \le r/4M$ for $0 \le t \le 1$. Let $\tau = \inf\{t > 0 :\ y(t) \text{ exists and } \|y(t)\|_{D_A(\theta+1,\infty)} \ge r/2M\}$. Since $\tau \ge 1$, then y is well defined and continuous at $t = \tau$ with values in $D_A(\theta+1,\infty)$, so that

$$r/2M = \|y(\tau)\|_{D_A(\theta+1,\infty)} \le M\|y_0\|_{D_A(\theta+1,\infty)}$$

$$+C_- \sup_{s\le\tau} \|g(x(s), y(s))\|_{D_A(\theta,\infty)} \le M\|y_0\|_{D_A(\theta+1,\infty)} + C_- M(r),$$

which is impossible if r is so small that $C_- M(r) \le r/4M$ and $\|y_0\|_{D_A(\theta+1,\infty)} \le r/4M$. For such values of r and of the norms of the initial data, $\|y(t)\|_{D_A(\theta+1,\infty)}$ remains bounded by $r/2M$, as far as it exists. ∎

We shall state the existence of a finite dimensional invariant manifold $\mathcal{V}_c$ for system (16), provided r is sufficiently small. Then we shall see that such a manifold attracts exponentially all the orbits starting from an initial datum sufficiently close to it. As a last step, we shall see that the null solution of (16) is stable, asymptotically stable, or unstable, if and only if it is stable, asymptotically stable, or unstable, with respect to the restriction of the flow to $\mathcal{V}_c$.

Theorem 10 *Let A satisfy (5) and (14). Then there exists $r_0 > 0$ such that for $r \le r_0$ there is a Lipschitz continuous function $\varphi : P(X) \mapsto (I - P)(D_A(\theta+1,\infty))$ such the graph of φ is invariant for system (16). If in* *addition F is k times continuously differentiable, then $\varphi \in C^{k-1}$, φ^{k-1} is Lipschitz continuous, and*

$$\varphi'(x)(A_0x + f(x,\varphi(x))) = A_-\varphi(x) + g(x,\varphi(x)), \quad x \in P(X). \tag{17}$$

The proof is quite similar to the one of Theorems 3.1 and 3.2 of [3], to which we refer the interested reader. Here we recall only what will be used later: as usual (see e.g. [5, Chapter 6] for the semilinear case), φ is sought as a fixed point of the operator Γ defined by

$$(\Gamma\varphi)(x) = \int_{-\infty}^{0} e^{-sA_-}g(z(s)),\varphi(z(s))ds, \quad x \in P(X), \tag{18}$$

where $z = z(s;x,\varphi)$ is the solution of the finite dimensional system

$$\begin{cases} z' = A_0z + f(z + \varphi(z)), \\ z(0) = x. \end{cases} \tag{19}$$

Fixed $a > 0$, one can show that Γ has a unique fixed point in the set

$$\begin{aligned} Y \;=\; &\{\varphi : P(X) \mapsto (I-P)(D_A(\theta+1,\infty)) : \varphi(0) = 0, \\ &\|\varphi(x)\|_{D_A(\theta+1,\infty)} \le r,\ \|\varphi(x) - \varphi(z)\|_{D_A(\theta+1,\infty)} \le a\|x - \overline{x}\|\}, \end{aligned}$$

provided r is small enough.

In the case where the fixed point of Γ is continuously differentiable, equality (17) follows by replacing $y = \varphi(x)$ in (16).

Let us prove a property of attractivity of the center manifold.

Proposition 11 *Let F be twice continuously differentiable. For every $\omega \in]0,\omega_-[$ there are $r(\omega)$, $M(\omega)$ such that if $\|x_0\|$, $\|y_0\|_{D_A(\theta+1,\infty)}$ are sufficiently small, the solution of (16) with $x(0) = x_0$, $y(0) = y_0$ exists in the large and satisfies*

$$\|y(t) - \varphi(x(t))\|_{D_A(\theta+1,\infty)} \le M(\omega)e^{-\omega t}\|y_0 - \varphi(x_0)\|_{D_A(\theta+1,\infty)}, \quad t \ge 0. \tag{20}$$

Proof — We use the notation and the results of Proposition 9. So, let $r \le r_0$ and $\|x_0\| + \|y_0\|_{D_A(\theta+1,\infty)} \le C_0(r)$, so that the solution of (16) exists in the large, and $\|y(t)\|_{D_A(\theta+1,\infty)} \le r$. Let $v(t) = y(t) - \varphi(x(t))$. Since φ satisfies (17), then

$$\begin{aligned} v'(t) &= A_-v(t) + g(x(t), y(t)) - g(x(t), \varphi(x(t))) \\ &\quad -\varphi'(x(t))(f(x(t), y(t)) - f(x(t), \varphi(x(t))) \\ &= A_-v(t) + G(x(t), y(t)), \quad t \ge 0. \end{aligned}$$

Recalling that $\|G(x(s), y(s))\|_{D_A(\theta,\infty)} \le L(r)(1+a)\|v(s)\|_{D_A(\theta+1,\infty)}$, from Proposition 4 with A replaced by A_-, we get

$$\sup_{0<t<T} e^{\omega t}\|v(t)\|_{D_A(\theta+1,\infty)}$$

$$\le C(\|v(0)\|_{D_A(\theta+1,\infty)} + L(r)(1+a) \sup_{0<t<T} e^{\omega t}\|v(t)\|_{D_A(\theta+1,\infty)}),$$

where $C = C(\omega)$. Taking r so small that $CL(r)(1+a) \le 1/2$, the statement follows, with $M(\omega) = 2C$. ∎

Using the technique of [3], one can see that the result of Proposition 11 holds under the mere assumption that F is continuously differentiable, but the proof is much longer.

Once exponential attractivity of $\mathcal{V}_c$ is established, one can prove that it is asymptotically stable with asymptotic phase, in the sense of next proposition. Again, the proof follows closely the proof of Theorem 3.3 of [3]; it is sufficient to replace everywhere $D_A(\theta)$ by $D_A(\theta,\infty)$ and $D_A(\theta+1)$ by $D_A(\theta+1,\infty)$.

Proposition 12 *For every $\omega \in]0, \omega_-[$ there is $C(\omega) > 0$ such that if $x_0 \in P(X)$ and $\|y_0\|_{D_A(\theta+1,\infty)} \le r_0$ then there exists $\overline{x} \in P(X)$, depending continuously on (x_0, y_0), such that, setting $\overline{z}(t) = z(t; \varphi, \overline{x})$, it holds*

$$\|x(t) - \overline{z}(t)\| + \|y(t) - \varphi(\overline{z}(t))\|_{D_A(\theta+1,\infty)} \le C(\omega)e^{-\omega t}\|y_0 - \varphi(x_0)\|, \quad t \ge 0. \tag{21}$$

Here r_0 is given by Proposition 9, and $z(t; \varphi, x)$ is the solution of (19).

From Proposition 12 we get easily the following corollary.

Corollary 13 *The null solution of (1) is stable (respectively, asymptotically stable, unstable) in $D_A(\theta+1,\infty)$ if and only if the null solution of the finite dimensional equation (19) is stable (respectively , asymptotically stable, unstable).*

4. Applications to nonlinear parabolic problems

We consider a nonlinear evolution problem in $[0,+\infty[\times\Omega$, Ω being a bounded open set in $\mathbf{R}^n$ with $C^{2+2\theta}$ boundary $\partial\Omega$, $0<\theta<1/2$:

$$\begin{cases} u_t = \Delta u + cu + f(u, Du, D^2u), & t \geq 0,\ x \in \overline{\Omega}, \\ \partial u(t,x)/\partial n = 0, & t \geq 0,\ x \in \partial\Omega, \end{cases} \tag{22}$$

$$u(0,x) = u_0(x), \quad x \in \Omega, \tag{23}$$

where Δ is the Laplace operator, $c \in \mathbf{R}$, f is a twice continuously differentiable function defined in a neighborhood of 0 in $\mathbf{R}\times\mathbf{R}^n\times\mathbf{R}^{n^2}$, with $f(0)=0$, $Df(0)=0$, and $\partial/\partial n$ denotes the normal derivative.

An example of an equation of the type (22) arising in Detonation Theory may be found in [2].

It is well known (see [12]) that the realization A of $\Delta + cI$ with homogeneous boundary condition in $X = C(\overline{\Omega})$ is sectorial, and that (see [1])

$$D_A(\theta,\infty) = C^{2\theta}(\overline{\Omega}),$$

with equivalence of the respective norms. The Schauder Theorem implies

$$D_A(\theta+1,\infty) = \{\psi \in C^{2\theta+2}(\overline{\Omega}) :\ \partial\psi/\partial n = 0\},$$

with equivalence of the norms. It follows that the mapping $\psi \mapsto F(\psi)$,

$$(F(\varphi))(x) = f(\psi(x), D\psi(x), D^2\psi(x)) \tag{24}$$

is well defined in a neighborhood of 0 in $D_A(\theta+1,\infty)$, with values in $D_A(\theta,\infty)$. Moreover, F is continuously differentiable, as it is easy to check.

Therefore, setting $u(t) = u(t,\cdot)$, problem (22) may be seen as an evolution equation in X of the type (1), such that the assumptions of Theorem 1 are satisfied. Applying Theorem 1, one proves local existence and uniqueness of a classical solution u of (22) for small initial data $u_0 \in C^{2+2\theta}(\overline{\Omega})$ such that $\partial u_0/\partial n = 0$. The regularity properties of the solution are the following: u_t and $D_{ij}u$ are continuous in $[0,T] \times \overline{\Omega}$, with $T > 0$, and they are 2θ-Hölder continuous with respect to the space variables, with Hölder constant independent of t; Δu is θ- Hölder continuous with respect to time, with Hölder constant independent of the space variables. Moreover, $t \mapsto u(t,\cdot)$ is continuous in $]0,T]$ with values in $C^{2\theta+2}(\overline{\Omega})$.

If we denote by $\{-\lambda_n\}_{n\in\mathbf{N}}$ is the ordered sequence of the eigenvalues of Δ with homogeneous Neumann boundary condition, then $\sigma(A) = \{-\lambda_n + c\}_{n\in\mathbf{N}}$. In the case where $c < 0$, the spectrum of A is contained in the left complex halfplane. Then the principle of linearized stability may be applied, and it gives existence in the large and exponential decay of the solution for small initial data. Precisely, since the first eigenvalue of A is c, then for every $\omega \in]0,c[$ there are $M(\omega)$, $R(\omega)$ such that if $\|u_0\|_{C^{2\theta+2}(\overline{\Omega})} \le R(\omega)$, then

$$\|u(t,\cdot)\|_{C^{2\theta+2}(\overline{\Omega})} \le M(\omega)e^{-\omega t}\|u_0\|_{C^{2\theta+2}(\overline{\Omega})}, \quad t \ge 0.$$

In the case where $c > 0$, the assumptions of Theorem 5 are satisfied, so that the null solution is unstable, and there exists a finite dimensional Lipschitz continuous unstable manifold. If $\lambda_n + c \neq 0$ for every $n \in \mathbf{N}$, then 0 is a saddle point: there exists also a Lipschitz continuous infinite dimensional stable manifold.

In the critical case of stability $c = 0$, the part of the spectrum of A in the imaginary axis consists only of the simple eigenvalue 0, and the corresponding eigenspace is the set of the constant functions. The operator A_0 is the null operator, and the projection P is given by

$$P\psi(\xi) = \frac{1}{meas\ \Omega}\int_\Omega \psi(y)dy, \quad \xi \in \overline{\Omega}.$$

If $f = f(p,q)$ does not depend explicitly on u, then F vanishes on $P(X)$, and this fact lets one decouple system (15), which becomes (with the notation of Section 3)

$$\begin{cases} x'(t) = PF(y(t)), \quad t \ge 0, \\ y'(t) = A_-y(t) + (I-P)F(y(t)), \quad t \ge 0. \end{cases} \tag{25}$$

The second equation satisfies the assumptions of the principle of linearized stability, with X replaced of course by $(I-P)(X)$, because the spectrum of A_- consists of the negative eigenvalues $-\lambda_n$, $n \geq 1$. By applying Theorem 3, we find that for every $\omega \in \,]0, \lambda_1[$ there are $R(\omega)$, $M(\omega)$ such that if $\|(I-P)u_0\|_{C^{2+2\theta}(\overline{\Omega})} \leq R(\omega)$ then the solution of the y-equation with initial value $y(0) = (I-P)u_0$ exists in the large and satisfies

$$\|y(t)\|_{C^{2+2\theta}(\overline{\Omega})} \leq M(\omega)e^{-\omega t}\|(I-P)u_0\|_{C^{2+2\theta}(\overline{\Omega})}, \quad t \geq 0.$$

Replacing in the x-equation, we find that $x(t)$ remains bounded, and

$$\|x(t)\|_{C^{2+2\theta}(\overline{\Omega})} = \|x(t)\|_{C(\overline{\Omega})} \leq \; const. \|u_0\|_{C^{2+2\theta}(\overline{\Omega})}, \quad t \geq 0.$$

Therefore, the null solution is stable in $C^{2+2\theta}(\overline{\Omega})$.

In the case where f depends explicitly on u, it is not easy to decouple the system as above, so we use the center manifold theory. Although it is very difficult, in general, to compute explicitly the function φ of Theorem 3, in our example we have $\varphi \equiv 0$, because F maps $P(X)$ into itself, so that for any r the unique small solution of (18) is the null function.

By Corollary 13, the null solution of (22) is stable (respectively, asymptotically stable, unstable) in $C^{2+2\theta}(\overline{\Omega})$ if and only if the null solution of the one dimensional equation

$$z'(t) = PF(z(t)), \quad t \geq 0, \tag{26}$$

is stable (respectively, asymptotically stable, unstable). Setting $z(t) = \zeta(t)\mathbf{1}$, where $\mathbf{1}$ is the constant function equal to 1 over $\overline{\Omega}$, (26) is equivalent to the scalar equation

$$\zeta'(t) = f(\zeta(t), 0, 0), \quad t \geq 0.$$

Assume that there is $k \in \mathbf{N}$ such that

$$a_k = \frac{\partial^k f}{\partial u^k}(0,0,0) \neq 0, \quad \frac{\partial^j f}{\partial u^j}(0,0,0) \;\; \text{for } j < k.$$

If either k is odd, or it is even and $a_k > 0$, then the null solution of (22) is unstable; if k is even and $a_k < 0$, the null solution of (22) is stable in $C^{2+2\theta}(\overline{\Omega})$.

Note that the instability part is trivial, since it is clear that the set of the constant functions is invariant for problem (22), i.e. if the initial datum u_0 is constant, then the solution depends only on t.

The same procedure may be performed in the case of the Dirichlet boundary condition. In the critical case of stability $c = \lambda_1$ (where $-\lambda_1$ is the first eigenvalue of Δ with Dirichlet boundary condition), things cannot be simplified as in the Neumann case because now $F(P(X))$ is not contained in $P(X)$, so that the function φ is not necessarily null. However, the computations made in [3] in the one dimensional case may be followed also for $n > 1$, replacing the little Hölder spaces by the corresponding Hölder spaces, without further modification. We refer to [3] for details.

References

1. P. Acquistapace, B. Terreni, *Hölder classes with boundary conditions as interpolation spaces*, Math. Z. **195** (1987), 451-471.

2. C.-M. Brauner, J. Buckmaster, J.W. Dold, C. Schmidt-Lainé, *On an evolution equation arising in detonation theory*, preprint.

3. G. Da Prato, A. Lunardi, *Stability, instability and center manifold theorem for fully nonlinear autonomous parabolic equations in Banach space*, Arch. Rat. Mech. Anal. **101** (1988), 115-141.

4. A.K. Drangeid, *The principle of linearized stability for quasilinear parabolic evolution equations* , Nonlinear Analysis T.M.A. **13** (1989), 1091-1113.

5. D. Henry, *Geometric theory of semilinear parabolic equations*, Lecture Notes in Math. **840**, Springer-Verlag, New York (1981).

6. A. Lunardi, *Asymptotic exponential stability in quasilinear parabolic equations*, Nonlinear Analysis T.M.A. **9** (1985), 563-586.

7. A. Lunardi, *Stability in fully nonlinear parabolic equations*, preprint.

8. A. Lunardi, *Analytic semigroups and optimal regularity in parabolic equations*, book in preparation.

9. A. Lunardi, *Bounded solutions of linear periodic abstract parabolic equations*, Proc. Royal Soc. Edinburgh **110A**, 135-159.

10. M. Poitier-Ferry, *The linearization principle for the stability of solutions of quasilinear parabolic equations, I*, Arch. Rat. Mech. Anal. **77** (1981), 301-320.

11. E. Sinestrari, *On the abstract Cauchy problem of parabolic type in spaces of continuous functions*, J. Math. Anal. Appl., **107**(1985), 16-66.

12. H.B. Stewart, *Generation of analytic semigroups by strongly elliptic operators under general boundary conditions*, Trans. Amer. Math. Soc. **259** (1980), 299-310.

Dipartimento di Matematica, Università di Cagliari
Via Ospedale 72, 09124 Cagliari, Italy

Fractional Integrals and Semigroups

Adam C. McBride

Introduction

In this paper we shall survey some of the connections between operators of fractional integration and semigroups of operators. For simplicity we shall restrict attention to operators related to the Riemann-Liouville fractional integral. On the one hand, such operators provide illustrations of the general theory of semigroups, particularly fractional power semigroups. On the other hand, it could be said that the operators have provided the stimulus for extensions of the general theory.

The paper is divided into four sections as follows.

1. Boundary values of holomorphic semigroups.
2. Fractional powers of certain operators mapping one space into a different space.
3. Fractional powers of certain operators mapping a space into itself.
4. α-times integrated semigroups.

1. For $\operatorname{Re}\alpha > 0$ and a suitable function ϕ we define $I^\alpha\phi$, *the Riemann-Liouville fractional integral of order α of ϕ*, by

$$(I^\alpha\phi)(x) = \frac{1}{\Gamma(\alpha)}\int_0^x (x-t)^{\alpha-1}\phi(t)dt. \tag{1}$$

G. R. Goldstein and J. A. Goldstein (eds.),
Semigroups of Linear and Nonlinear Operations and Applications, 205–224.

When $\alpha = n$, we obtain an operator corresponding to repeated (n-fold) integration.

First consider the properties of I^α relative to the spaces $L^p(0,1)$ so that we take $0 < x < 1$ in (1). From [9, pp. 664 et seq.] we know that $\{I^\alpha : \operatorname{Re}\alpha > 0\}$ gives rise to a holomorphic semigroup of bounded linear operators on $L^p(0,1)$ (for $1 \le p \le \infty$) and *the first index law*

$$I^\alpha I^\beta = I^{\alpha+\beta} = I^\beta I^\alpha \quad (\operatorname{Re}\alpha > 0,\ \operatorname{Re}\beta > 0) \tag{2}$$

holds in the sense of operators on $L^p(0,1)$.

On writing $\alpha = \mu + i\nu(\mu, \nu$ real with $\mu > 0)$, we may consider the behaviour of I^α as $\mu = \operatorname{Re}\alpha \to 0+$. For $1 < p < \infty$, it was proved by Kalisch [13] and Fisher [6] that, for each fixed $\phi \in L^p(0,1)$, the limit

$$I^{i\nu}\phi = \lim_{\mu \to 0+} I^{\mu+i\nu}\phi \tag{3}$$

was well-defined, the limit existing in the $L^p(0,1)$ norm. Furthermore, the family of operators $\{I^{i\nu} : \nu \in \mathbf{R}\}$ so defined forms a strongly continuous group of bounded operators on $L^p(0,1)$. We can think of this group as giving the *boundary values* of the original holomorphic semigroup and, as one might expect from (2),

$$I^\alpha I^{i\nu} = I^{\alpha+i\nu} = I^{i\nu} I^\alpha \quad (\operatorname{Re}\alpha > 0,\ \nu \in \mathbf{R}) \tag{4}$$

in the sense of operators on $L^p(0,1)$ for $1 < p < \infty$.

Now let us replace $L^p(0,1)$ by $L^p(0,\infty)$ where $1 < p < \infty$. The operators I^α $(\operatorname{Re}\alpha > 0)$ are now unbounded operators whose domains are proper subspaces of $L^p(0,\infty)$. Nevertheless, Fisher [6] proved that it was still possible to obtain a *boundary group*

$\{I^{i\nu} : \nu \in \mathbf{R}\}$ of bounded linear operators on the whole space $L^p(0,\infty)$, although the limit process was more elaborate than that in (3). Equation (4) remains valid in the sense that

$$I^\alpha I^{i\nu}\phi = I^{\alpha+i\nu}\phi = I^{i\nu}I^\alpha\phi \quad (\mathrm{Re}\,\alpha > 0,\ \nu \in \mathbf{R})$$

whenever

$$\phi \in D(I^\alpha) = \{\phi \in L^p(0,\infty) : I^\alpha\phi \in L^p(0,\infty)\}.$$

These results have led to more general investigations into boundary values of holomorphic semigroups of (possibly) unbounded linear operators in a Banach space X. We mention in particular the work of Hughes and Kantorovitz [12] who introduced the concept of a *regular semigroup of operators* and showed that such semigroups $\{T(\alpha) : \mathrm{Re}\,\alpha > 0\}$ gave rise to boundary groups $\{T(i\nu) : \nu \in \mathbf{R}\}$ of bounded linear operators, the boundary groups being strongly continuous. As indicated in [12], this is one instance where results for fractional integrals have suggested an extension of the general theory of semigroups.

2. We have already mentioned that, in the setting of $L^p(0,\infty)$, the Riemann-Liouville fractional integral is an unbounded operator. One way to remedy the situation is to introduce weighted spaces with simple powers as weights. In this section and the next we shall work within the framework of the Banach spaces $L_{p,\mu}$ $(1 < p < \infty, \mu \in \mathbf{C})$ where

$$L_{p,\mu} = \{\phi : \| \phi \|_{p,\mu} < \infty\}$$
$$\| \phi \|_{p,\mu} = \{\int_0^\infty | x^{-\mu}\phi(x) |^p \, dx/x\}^{\frac{1}{p}}. \tag{5}$$

It is clear that $L_{p,\mu}$ is homeomorphic to the usual space $L^p(0,\infty)$ under the mapping $\phi(x) \to x^{-\mu-1/p}\phi(x)$. (We shall exclude the cases $p = 1$ and $p = \infty$ although many of our results apply in such spaces also.)

If we take $\phi(x) = x^\lambda$ with $\operatorname{Re}\lambda > -1$, then

$$I^\alpha\phi(x) = \frac{1}{\Gamma(\alpha)}\int_0^x (x-t)^{\alpha-1}t^\lambda dt = \frac{\Gamma(\lambda+1)}{\Gamma(\lambda+\alpha+1)}x^{\lambda+\alpha}$$

i.e.

$$I^\alpha x^\lambda = \frac{\Gamma(\lambda+1)}{\Gamma(\lambda+\alpha+1)}x^{\lambda+\alpha}. \tag{6}$$

The change in power from λ to $\lambda+\alpha$ can easily be accommodated within the structure of our weighted spaces. It can be shown that

I^α is a continuous linear mapping from $L_{p,\mu}$ into $L_{p,\mu+\alpha}$ provided that $\operatorname{Re}\mu > -1, \operatorname{Re}\alpha > 0.$ (7)

Thus each I^α (and I^1 in particular) maps from one weighted space into a *different* one.

We shall now consider a class of operators, each of which maps from one $L_{p,\mu}$ space into a different one. For each such operator T we shall define a general power T^α ($\operatorname{Re}\alpha > 0$) and thereby obtain a fractional power semigroup. The Riemann-Liouville fractional integral can be recovered as a special case. However, by restricting attention to smooth functions, we can also treat fractional derivatives and powers of "Bessel type" differential operators. We shall merely outline the theory here. Full details can be found in [21], [22] and [23], with an edited version in [24, pp. 99-139].

We shall make extensive use of the Mellin transform $\mathcal{M}$ defined formally by

$$(\mathcal{M}\phi)(s) = \int_0^\infty x^{s-1}\phi(x)dx. \tag{8}$$

(For $\phi \in L_{p,\mu}$, the integral exists via mean convergence provided that $1 < p \leq 2$ and $\operatorname{Re} s = -\operatorname{Re} \mu$.) The Mellin convolution $k * \phi$ of two functions k and ϕ is defined by

$$(k * \phi)(x) = \int_0^\infty k(\frac{x}{t})\phi(t)\frac{dt}{t} \quad (x > 0). \tag{9}$$

Under appropriate conditions, we obtain

$$(\mathcal{M}(k * \phi))(s) = (\mathcal{M}k)(s)(\mathcal{M}\phi)(s). \tag{10}$$

For fixed k, we can think of the transform S defined by

$$S\phi = k * \phi. \tag{11}$$

It follows from (10) and (11) that

$$(\mathcal{M}(S\phi))(s) = (\mathcal{M}k)(s)(\mathcal{M}\phi)(s). \tag{12}$$

It is then possible to obtain the mapping properties of the transform S by studying its multiplier $\mathcal{M}k$ and extensive investigations have been carried out by Rooney [26], [27]. (A change of variable relates this to symbols of pseudo-differential operators defined via the Fourier transform.) Under appropriate conditions on $\mathcal{M}k$, S will map $L_{p,\mu}$ into itself and we then call $\mathcal{M}k$ *an* $L_{p,\mu}$ *multiplier.*

Now let γ be a non-zero complex number. Let

$$T = x^{-\gamma} S \tag{13}$$

where S is as in (11). (This means $(T\phi)(x) = x^{-\gamma}(S\phi)(x)$.) Then from (12) and (13) we get

$$(\mathcal{M}(T\phi))(s - \gamma) = (\mathcal{M}k)(s)(\mathcal{M}\phi)(s). \tag{14}$$

We now assume that the multiplier $\mathcal{M}k$ has a particular 'factorised" form, namely $h(s-\gamma)/h(s)$ for some function h. Thus we finally arrive at operators T which satisfy an equation of the form

$$(\mathcal{M}(T\phi))(s-\gamma)=\frac{h(s-\gamma)}{h(s)}(\mathcal{M}\phi)(s). \tag{15}$$

A simple induction based on (15) gives formally

$$(\mathcal{M}(T^n\phi))(s-n\gamma)=\frac{h(s-n\gamma)}{h(s)}(\mathcal{M}\phi)(s) \quad (n=1,2,\cdots)$$

and this immediately suggests that an operator T^α can be defined by requiring that

$$(\mathcal{M}(T^\alpha\phi))(s-\alpha\gamma)=\frac{h(s-\alpha\gamma)}{h(s)}(\mathcal{M}\phi)(s). \tag{16}$$

This will work provided that $h(s-\alpha\gamma)/h(s)$ is an acceptable $L_{p,\mu}$ multiplier, in which case T^α will be a continuous linear mapping from $L_{p,\mu}$ into $L_{p,\mu+\alpha\gamma}$ under appropriate conditions. We may feel justified in referring to T^α as *an αth power of* T.

It is worth remarking that a general αth power defined in this way is not unique. To see this we note that if (15) holds for a particular function h then it will also hold for each of the functions $h_r (r=1,2,\cdots)$ where

$$h_r(s)=\exp(2r\pi is/\gamma)h(s). \tag{17}$$

However, as regards (16),

$$\frac{h_r(s-\alpha\gamma)}{h_r(s)}=\exp(-2r\pi i\alpha)\frac{h(s-\alpha\gamma)}{h(s)}$$

and, since $\exp(-2r\pi i\alpha)\neq 1$ in general, we may obtain infinitely many possibilities for T^α.

Suppose now that T satisfies (15) and that we use the same choice of h throughout (i.e. for the calculation of all powers). Then it is routine to check the validity of *the first index law*

$$T^{\alpha}T^{\beta} = T^{\alpha+\beta} = T^{\beta}T^{\alpha} \tag{18}$$

under appropriate conditions. Perhaps more surprisingly it is also easy to deal with *the second index law*

$$(T^{\alpha})^{\beta} = T^{\alpha\beta}. \tag{19}$$

Indeed, with the same h throughout we calculate $(T^{\alpha})^{\beta}$ by replacing T, α and γ by T^{α}, β and $\alpha\gamma$ respectively in (16). This produces (16) with α replaced by $\alpha\beta$ and (19) is proved formally. The simplicity contrasts with the difficulties encountered in the spectral approach which is introduced in the next section.

It is time to illustrate our theory and, as promised, we shall show first how to recover I^{α}. All we need to do here (and in subsequent cases) is to identify γ and a suitable h.

We start with $T = I^1$ where

$$(I^1\phi)(x) = \int_0^x \phi(t)dt$$

and find that (under appropriate conditions, e.g. for $\phi \in C_0^{\infty}(0,\infty)$)

$$(\mathcal{M}(I^1\phi))(s-1) = \frac{1}{1-s}(\mathcal{M}\phi)(s).$$

Comparison with (15) gives $\gamma = 1$ and a suitable choice for h is

$$h(s) = \frac{1}{\Gamma(1-s)}. \tag{20}$$

Formula (16) then says that $I^\alpha \equiv (I^1)^\alpha$ has to be such that

$$(\mathcal{M}(I^\alpha \phi))(s-\alpha) = \frac{\Gamma(1-s)}{\Gamma(1-s+\alpha)}(\mathcal{M}\phi)(s). \tag{21}$$

We can check directly that this gives the operator in (1) so that the choice of h in (20) is in a sense canonical.

To discuss derivatives we need to use smooth functions. With $D \equiv \dfrac{d}{dx}$ and $\phi \in C_0^\infty(0,\infty)$, we may integrate by parts to obtain

$$(\mathcal{M}(D\phi))(s+1) = -s(\mathcal{M}\phi)(s).$$

Comparison with (15) gives $\gamma = -1$ and a suitable choice for h is given once again by (20). Then D^α has to be such that

$$(\mathcal{M}(D^\alpha \phi))(s+\alpha) = \frac{\Gamma(1-s)}{\Gamma(1-s-\alpha)}(\mathcal{M}\phi)(s). \tag{22}$$

Examination of (21) and (22) leads us to conclude formally that

$$I^\alpha = D^{-\alpha}; \quad D^\alpha = I^{-\alpha} \tag{23}$$

as one might expect.

It is worth commenting briefly on the validity of our formal calculations. The multiplier $\Gamma(1-s)/\Gamma(1-s+\alpha)$ in (21) can serve as an $L_{p,\mu}$ multiplier if $\operatorname{Re}\alpha > 0$ (or if $\alpha = 0$, in which case we obtain the identity operator) but not if $\operatorname{Re}\alpha < 0$. When we restrict attention to smooth functions in a subspace $F_{p,\mu}$ of $L_{p,\mu}$ (see Definition 6.1 in [24], for instance), we discover that (21) can serve as an $F_{p,\mu}$ multiplier for any α and can then proceed to establish (22) and (23). Again see [24] for further details.

Sometimes we may wish to differentiate or integrate with respect to a positive power x^m of the variable rather than x itself.

It is possible to define operators I_m^α and D_m^α which extend the previous results for $m = 1$. Thus, in $L_{p,\mu}$ with $\operatorname{Re}\alpha > 0$, (21) becomes

$$(\mathcal{M}(I_m^\alpha\phi))(s - m\alpha) = \frac{\Gamma(1-\frac{s}{m})}{\Gamma(i-\frac{s}{m}+\alpha)}(\mathcal{M}\phi)(s). \qquad (24)$$

Such generalisations are important in connection with special functions.

One way of justifying the last remark is to consider an n-th order differential operator of the general form

$$T = x^{a_1}Dx^{a_2}D\cdots x^{a_n}Dx^{a_{n+1}}. \qquad (25)$$

As an example, consider

$$\begin{aligned} B_\nu &= x^{-\nu-1}Dx^{2\nu+1}Dx^{-\nu} \\ &= x^{-2}(x^2D^2 + xD - \nu^2). \end{aligned} \qquad (26)$$

The equation $B_\nu y = -y$ is *Bessel's equation of order* ν. Consequently the operator (26) is sometimes called *a hyper-Bessel operator*. To define T^α via (16), we assume that $\sum_{j=1}^{n+1} a_j$ is real and that

$$m = \left| \sum_{j=1}^{n+1} a_j - n \right| > 0. \qquad (27)$$

(If the modulus is zero we are in the situation of the next section.) On replacing each D in (25) by $mx^{m-1}\frac{d}{dx^m}$, with m as in (27), a concrete expression can be obtained. See [20] and [24], these papers having been motivated by Sprinkhuizen-Kuyper [29]. Notice that for B_ν in (26) we get $m = 2$. This explains the appearance of fractional integrals with respect to x^2 in formulae for the Bessel functions such as Sonine's integral [5, §7.12].

To summarise our theory, we may say that the formula (15) leads to a very simple method of defining T^α and hence of obtaining a fractional power semigroup (or even a group). The fact that T maps between different spaces means that some features are missing by comparison with what will follow in the next section. For example, since $\lambda I - T$ is meaningless, we do not have resolvent operators available. Analyticity with respect to α has to be handled at a lower level, with Fréchet derivatives unavailable. Thus we are forced to fix not only $\phi \in L_{p,\mu}$ but also $x \in (0, \infty)$ and then to investigate $(T^\alpha \phi)(x)$ as a function of α. Our final comment here is that the above results are thirled to the Mellin transform and the weighted spaces $L_{p,\mu}$. In contrast we shall now turn to the spectral approach for operators mapping a general Banach space into itself.

3. We now consider operators mapping a space (or a subspace of it) into the same space. Precisely, let X be a Banach space and let A be a linear operator whose domain $D(A)$ and range $R(A)$ are linear subspaces of X. We shall review the basic method of defining powers of $-A$. See [3], [11] and [17].

First recall that, if A is a positive real number and α is a complex number satisfying $0 < \operatorname{Re} \alpha < 1$, then

$$A^{\alpha-1} = \frac{\sin \pi\alpha}{\pi} \int_0^\infty \frac{\lambda^{\alpha-1}}{\lambda + A} d\lambda. \tag{28}$$

The formal analogue of (28) for operators is

$$(-A)^{\alpha-1} = \frac{\sin \pi\alpha}{\pi} \int_0^\infty \lambda^{\alpha-1}(\lambda I - A)^{-1} d\lambda$$

or

$$(-A)^\alpha x = \frac{\sin \pi\alpha}{\pi} \int_0^\infty \lambda^{\alpha-1} R(\lambda, A)(-A) x \, d\lambda \tag{29}$$

where $R(\lambda, A) = (\lambda I - A)^{-1}$ and $0 < \text{Re}\,\alpha < 1$ as before. For (29) to exist we certainly need $x \in D(A)$ as well as requiring $R(\lambda, A)$ to exist for all $\lambda > 0$. To guarantee convergence of (29) as a Bochner integral we assume that

$$\begin{aligned} &(0, \infty) \subset \rho(A), \text{ the resolvent set of } A \\ &\| \lambda R(\lambda, A) \| \leq M \text{ for all } \lambda > 0 \end{aligned} \tag{30}$$

where M is a positive constant.

We observe that condition (30) is satisfied when A is the infinitesimal generator of a uniformly bounded C_0-semigroup, as a consequence of the Hille-Yosida Theorem.

In order to define $(-A)^\alpha$ for $\text{Re}\,\alpha > 0$ rather than for the restricted range $0 < \text{Re}\,\alpha < 1$, we first observe that (29) can be rewritten in the form

$$\begin{aligned} (-A)^\alpha x = {} & \frac{\sin \pi\alpha}{\pi} \int_0^\infty \lambda^{\alpha-1} [R(\lambda, A) - \frac{\lambda}{1+\lambda^2}](-A) x \, d\lambda \\ & + \sin \frac{\pi\alpha}{2} (-A) x. \end{aligned} \tag{31}$$

However the expression (31) is meaningful for the larger range of values $0 < \text{Re}\,\alpha < 2$. (Basically $R(\lambda, A) - \frac{\lambda}{1+\lambda^2} I$ behaves like λ^{-2} as $\lambda \to \infty$.) We can therefore use (31) to extend the definition of $(-A)^\alpha x$ to this larger range for $x \in D(A)$. Finally, if α satisfies $n - 1 < \text{Re}\,\alpha < n + 1$ for a positive integer n, we define $(-A)^\alpha$ via

$$(-A)^\alpha x = (-A)^{\alpha-n+1} (-A)^{n-1} x \tag{32}$$

for suitable x. If for simplicity we assume that $D(A) = X$ and that A is bounded, then the family $\{(-A)^\alpha : \text{Re}\,\alpha > 0\}$ is a holomorphic semigroup.

To illustrate this theory we return to our theme of the Riemann-Liouville fractional integral. In considering (6) in the previous section, we introduced a family of different spaces. Alternatively we can modify I^α by considering the operator $I^{\eta,\alpha}$ defined by

$$(I^{\eta,\alpha}\phi)(x) = x^{-\eta-\alpha}I^\alpha x^\eta \phi(x) = \frac{x^{-\eta-\alpha}}{\Gamma(\alpha)}\int_0^x (x-t)^{\alpha-1}t^\eta\phi(t)dt. \tag{33}$$

In contrast to (6) we find that, for $\mathrm{Re}\,\alpha > 0$ and $\mathrm{Re}(\eta+\lambda) > -1$

$$I^{\eta,\alpha}x^\lambda = \frac{\Gamma(\eta+\lambda+1)}{\Gamma(\eta+\alpha+\lambda+1)}x^\lambda. \tag{34}$$

In view of (34) it is no surprise that

$I^{\eta,\alpha}$ is a continuous linear mapping from $L_{p,\mu}$ into itself if $\mathrm{Re}\,\alpha > 0,\ \mathrm{Re}(\eta+\mu) > -1.$ (35)

The operator (33) is an example of an Erdélyi-Kober operator. Such operators were studied by Erdélyi and Kober in a series of papers [4], [15] and [16]. Modifications involving fractional integrals with respect to x^2 rather than x subsequently led to an elegant method for solving dual integral equations arising in potential theory. (This is related to the emergence of $m = 2$ in connection with the Bessel operator (26).) For further details see the article by Sneddon in [28] as well as Chapter 7 in [19].

Consider the operator $I^{\eta,1}$ on $L^p(0,\infty)$ for simplicity. (The results go through for any $L_{p,\mu}$ space with minor changes.) Take $A = -I^{\eta,1}$ in order to define powers of $I^{\eta,1}$. The results which follow are due to Lamb [18]. As regards (30) we find that, for $\lambda > 0$,

$$R(\lambda, -I^{\eta,1}) = \frac{1}{\lambda}I - \frac{1}{\lambda^2}I^{\eta+\frac{1}{\lambda},1} \tag{36}$$

provided that $\mathrm{Re}\,\eta > -1$ (where the first I on the right-hand side denotes the identity operator!). We remark that (36) can be checked by showing that both sides have the same Mellin multiplier in the sense of (12), namely $(\eta + 1 - s)/(\lambda\eta + \lambda + 1 - \lambda s)$. We then obtain

$$\begin{aligned} &\| \lambda R(\lambda, -I^{\eta,1}) \| = \| I - \frac{1}{\lambda} I^{\eta+\frac{1}{\lambda},1} \| \\ &\leq 1 + \frac{1}{1 + \lambda(\eta + 1)} \leq 1 + \frac{1}{1} = 2 \text{ for } \mathrm{Re}\,\lambda > 0,\ \mathrm{Re}\,\eta > -1. \end{aligned}$$

This completes the verification of (30). On substituting (36) into (31) and then using (32) we find that, for $\mathrm{Re}\,\alpha > 0$ and $\mathrm{Re}\,\eta > -1$,

$$(I^{\eta,1})^{\alpha} = x^{-(\eta+1)} H^{\alpha} x^{\eta+1}$$

where the operator H^{α} is defined by

$$(H^{\alpha}\phi)(x) = \frac{1}{\Gamma(\alpha)} \int_0^x [\log(\frac{x}{t})]^{\alpha-1} \phi(t) \frac{dt}{t}. \tag{37}$$

The operator H^{α} in (37) corresponds to integrating α times with respect to $\log x$. It is often linked with the name of Hadamard.

In the above example, everything could be calculated explicitly. However the case of $I^{\eta,2}$ presents more difficulty and the case of $(I^{\eta,\alpha})^{\beta}$ for general α and β seems hopeless.

4. In this final section we shall consider families of bounded operators obtained as fractional integrals of semigroups. Consider a C_0-semigroup $\{T(t)\}_{t\geq 0}$ of bounded linear operators on a Banach space X and let $A : D(A) \to X$ be its infinitesimal generator. Then the abstract Cauchy problem (ACP)

$$\frac{du}{dt} = Au \quad (t > 0);\ u(0) = u_0 \tag{38}$$

has a unique "classical" solution $u : [0,\infty) \to X$ for any given $u_0 \in D(A)$ and u is given by

$$u(t) = T(t)u_0 \quad (t \geq 0).$$

It is of interest to ask what happens if A does not generate a C_0-semigroup.

Let α be real and positive and let $\{T(t)\}_{t\geq 0}$ be as above. We define a new family $\{S(s)\}_{s\geq 0}$ of bounded linear operators on X by

$$S(s) = \frac{1}{\Gamma(\alpha)} \int_0^s (s-t)^{\alpha-1} T(t)dt \tag{39}$$

the integral converging with respect to the operator norm on $B(X)$. We shall refer to the family $\{S(s)\}_{s\geq 0}$ as an α-times integrated semigroup.

This concept was introduced in the case $\alpha = 1$ by Arendt [1] and thereafter the theory has been extended to positive integers α and finally to all positive values of α. Details can be found in the papers of Kellermann and Hieber [14], Neubrander [25], Thieme [30] and Hieber [7], [8]. Basically (39) is an operator version of (1) and we could express this briefly in the form

$$S = I^\alpha T. \tag{40}$$

If A denotes the infinitesimal generator of $\{T(t)\}$, we may apply the convolution theorem for the (operator-valued) Laplace transform to deduce from (39) that

$$\int_0^\infty e^{-\lambda s} S(s)ds = \frac{1}{\lambda^\alpha} R(\lambda, A)$$

i.e.

$$R(\lambda, A) = \lambda^\alpha \int_0^\infty e^{-\lambda s} S(s)ds \tag{41}$$

for all sufficiently large (real) λ. It is evident that A not only generates $\{T(t)\}_{t\geq 0}$ but also generates $\{S(s)\}_{s\geq 0}$ in a sense embodied in (41). However, it is possible to find operators A which will "generate" a family $\{S(s)\}_{s\geq 0}$ satisfying (41) without generating a C_0-semigroup. We shall use the following definition.

Definition A linear operator $A : D(A) \to X$ is said to be the generator of an α-times integrated semigroup (for $\alpha \geq 0$) if

(i) the resolvent set, $\rho(A)$, of A contains (ω, ∞) for some $\omega \in \mathbf{R}$

(ii) there exists a mapping $S : [0, \infty) \to X$ which is strongly continuous and satisfies

$$\| S(s) \| \leq M e^{\omega s} \quad (s \geq 0)$$

(where M is a positive constant) and

$$R(\lambda, A) = \lambda^{\alpha} \int_0^{\infty} e^{-\lambda s} S(s) ds$$

for $\lambda > \max(\omega, 0)$.

For $\alpha = 0$, the above definition coincides with the usual infinitesimal generator in view of the Hille-Yosida Theorem. A "Hille-Yosida Theorem" can be proved for $\alpha > 0$ too and the theory in this case bears some similarities with that for $\alpha = 0$. For example $\{S(s)\}$ is uniquely determined by A. There are also some differences, notably that

$$\textit{for } \alpha > 0, D(A) \textit{ need not be dense in } X. \tag{42}$$

A useful fact is that

if A generates an α-times integrated semigroup, then A generates a β-times integrated semigroup for all $\beta > \alpha$.

This means that we can accommodate more operators as generators by increasing α.

With this in mind let us return to the ACP (38) and suppose that A generates an r-times integrated semigroup for some non-negative integer r. It is well known (see, for instance [2]) that (38) will have a unique "classical" solution for each $u_0 \in D(A^{r+1})$. This has been extended to positive non-integral values of α by Hieber [7].

Theorem Let $\alpha \geq 0$, $\epsilon > 0$ and assume that A generates an α-times integrated semigroup $\{S(s)\}_{s \geq 0}$ satisfying

$$\| S(s) \| \leq M s^{\alpha} e^{\omega s} \quad (s \geq 0)$$

for non-negative constants M, ω. Then there exists a unique classical solution of (38) for all $u_0 \in D((-A)^{\alpha+\epsilon+1})$.

This result indicates a connection with the theory of fractional powers of operators discussed in the previous section. At this juncture, it is legitimate to ask if there are any important applications which require the use of a non-integral value of α, thereby justifying the use of "fractional" integration. To provide this justification, we mention another result of Hieber.

Consider the ACP (38) for the Schrödinger equation

$$\frac{du}{dt} = i\Delta u \quad (t > 0); \quad u(0) = u_0 \tag{43}$$

in the Banach space $L^p(\mathbf{R}^n)$ where we take $1 < p < \infty$ for simplicity. Hörmander [10] proved that (when defined on its natural domain) $i\Delta$ generates a C_0-semigroup on $L^p(\mathbf{R}^n)$ if and only if $p = 2$. In contrast Hieber [8] has shown that $i\Delta$ generates an

α-times integrated semigroup on $L^p(\mathbf{R}^n)$ if and only if

$$\alpha \geq n \left| \frac{1}{2} - \frac{1}{p} \right| . \tag{44}$$

(This confirms that $\alpha = 0$ is only possible when $p = 2$.) It is of interest to investigate the space of initial conditions u_0 for which (43) has a unique classical solution u. The "optimal" space turns out to be the Sobolev space $W^{n+2,p}$. To obtain this when n is odd it is necessary to use fractional values of α, as use of r-times integrated semigroups with r a non-negative integer will give a weaker result which only guarantees existence and uniqueness of a classical solution for $u_0 \in W^{n+3,p}$.

Conclusion This survey paper has touched on a few of the interconnections between the Riemann-Liouville fractional integrals, fractional powers of operators and semigroups of operators. It may be expected that all three areas will continue to play a role in the future study of evolution equations and abstract Cauchy problems.

References

1. W. Arendt. Resolvent positive operators, *Proc. London Math. Soc.* 54: 321-349 (1987).

2. W. Arendt. Vector-valued Laplace transforms and Cauchy problems, *Israel J. Math.* 59: 327-352 (1987).

3. A. V. Balakrishnan. Fractional powers of closed operators and semigroups generated by them, *Pacific J. Math.* 10: 419-437 (1960).

4 A Erdélyi. On fractional integration and its application to the theory of Hankel transforms, *Quart. J. Math. (Oxford)* 11: 293-303 (1940).

5. A. Erdélyi, W. Magnus, F. Oberhettinger and F. Tricomi. *Higher Transcendental Functions*, Vol. 2. McGraw-Hill, New York (1953).

6. M. J. Fisher. Imaginary powers of the indefinite integral, *Amer. J. Math.* 93: 317-328 (1971).

7. M. Hieber. Integrated semigroups and differential operators on L^p-spaces, *Math. Ann.* 291: 1-16 (1991).

8. M. Hieber. Laplace transforms and α-times integrated semigroups, *Forum Math.* 3: 595-612 (1991).

9. E . Hille and R. S. Phillips. *Functional Analysis and Semi-groups*. American Mathematical Society, Providence (1957).

10. L. Hörmander. Estimates for translation invariant operators in L^p spaces, *Acta Math.* 104: 93-139 (1960).

11. H. W. Hövel and U. Westphal. Fractional powers of closed operators, *Studia Math.* 42: 177-194 (1972).

12. R. J. Hughes and S. Kantorovitz. Boundary values of holomorphic semigroups of unbounded operators and similarity of certain perturbations, *J. Funct. Anal.* 29: 253-273 (1978).

13. G. K. Kalisch. On fractional integrals of purely imaginary order in L^p, *Proc. Amer. Math. Soc.* 18: 136-139 (1967).

14. H. Kellermann and M. Hieber. Integrated semigroups, *J. Funct. Anal.* 84: 160-180 (1989).

15. H. Kober. On fractional integrals and derivatives, *Quart. J. Math. (Oxford)* 11: 193-211 (1940).

16. H. Kober and A. Erdélyi. Some remarks on Hankel transforms, *Quart. J. Math. (Oxford)* 11: 212-221 (1940).

17. H. Komatsu. Fractional powers of operators, *Pacific J. Math.* 19: 285-346 (1966).

18. W. Lamb. A spectral approach to an integral equation, *Glasgow Math. J.* 26: 83-89 (1985).

19. A. C. McBride. *Fractional Calculus and Integral Transforms of Generalized Functions*. Pitman, London (1979).

20. A. C. McBride. Fractional powers of a class of ordinary differential operators, *Proc. London Math. Soc.* (3) 45: 519-546 (1982).

21. A. C. McBride. Fractional powers of a class of Mellin multiplier transforms I, *Appl. Anal.* 21: 89-127 (1986).

22. A. C. McBride. Fractional powers of a class of Mellin multiplier transforms II, *Appl. Anal.* 21: 129-149 (1986).

23. A. C. McBride. Fractional powers of a class of Mellin multiplier transforms III, *Appl. Anal.* 21: 151-173 (1986).

24. A. C. McBride and G. F. Roach (eds.). *Fractional Calculus*. Pitman, London (1985).

25. F. Neubrander. Integrated semigroups and their applications to the abstract Cauchy problem, *Pacific J. Math.* 135: 111-155 (1988).

26. P. G Rooney. On the ranges of certain fractional integrals, *Canad. J. Math.* 24: 1198-1216 (1972).

27. P. G. Rooney. A technique for studying the boundedness and extendability of certain types of operators, *Canad. J. Math.* 25: 1090-1102 (1973).

28. B. Ross (ed.). *Fractional Calculus and its Applications.* Springer-Verlag, Berlin (1975).

29. I. G. Sprinkhuizen-Kuyper. A fractional integral operator corresponding to negative powers of a certain second order differential operator, *J. Math. Anal. Appl.* 72: 674-702 (1979).

30. H. R. Thieme. "Integrated semigroups" and integrated solutions to abstract Cauchy problems, *J. Math. Anal. Appl.* 152: 416-447 (1990).

Department of Mathematics

University of Strathclyde

Livingstone Tower

26 Richmond Street

Glasgow G1 1XH

SCOTLAND, U.K.

Spectral and asymptotic properties of strongly continuous semigroups

Rainer Nagel

Abstract: The characterization of the asymptotic behavior of strongly continuous semigroups through spectral properties of its generator is a classical subject. Recently some important progress has been made which will be surveyed in this paper.

Section 1

Some examples

The asymptotic behavior of strongly continuous semigroups $(T(t))_{t \geq 0}$ of linear operators on Banach spaces is intimately related to spectral properties. In fact, information on the spectrum $\sigma(A)$ of the generator $(A, D(A))$ is in many situations sufficient in order to describe the qualitative behavior of the operators $T(t)$ as $t \to \infty$. We are mainly interested in the existence of

$$P := \lim_{t \to \infty} T(t),$$

or, more generally, in the relative compactness of $\{T(t) : t \geq 0\}$ for some appropriate topology. Before starting the systematic discussion we present various examples from different areas of analysis.

First, let $A \in M_n(\mathbb{C})$ be a complex $(n \times n)$-matrix and consider the semigroup $(e^{tA})_{t \geq 0}$ generated by A. The exponential stability of this

G. R. Goldstein and J. A. Goldstein (eds.),
Semigroups of Linear and Nonlinear Operations and Applications, 225–240.

semigroup is characterized through spectral properties in the following classical

1.1 Liapunov Stability Theorem. *For a matrix $A \in M_n(\mathbb{C})$ and the semigroup $(e^{tA})_{t\geq 0}$ the following assertions are equivalent.*

(a) $\lim_{t\to\infty} \|e^{tA}\| = 0$.

(b) $\lim_{t\to\infty} e^{\epsilon t}\|e^{tA}\| = 0$ *for some* $\epsilon > 0$.

(c) *The spectral bound* $s(A) := sup\{\mathrm{Re}\,\lambda : \lambda \in \sigma(A)\}$ *satisfies* $s(A) < 0$.

(d) *There exists a positive semidefinite matrix* $R \in M_n(\mathbb{C})$ *such that* $RA + A^*R = -Id$.

For the proof we refer to [H-J] Thm. 2.2.1, but we point out that condition (d) is in fact a "spectral" condition: To see this one looks at the operator Δ_A defined on the space $M_n(\mathbb{C})$ by

$$\Delta_A(T) := TA + A^*T \quad \text{for } T \in M_n(\mathbb{C}).$$

Then it can be shown that (d) is equivalent to
(d') 0 is in the resolvent set of Δ_A and the resolvent $R(0, \Delta_A) := -\Delta_A^{-1}$ maps the cone of positive semidefinite matrices into itself.

Next we consider a strictly positive probability measure μ on $\mathbb{R}$ and the operator

$$Mf(x) := ix \cdot f(x), \quad x \in \mathbb{R},$$

for $f \in L^p(\mathbb{R}, \mu)$, $1 \leq p < \infty$. Then M with domain $D(M) := \{f \in L^p : Mf \in L^p\}$ generates a group of isometries

$$T(t)f(x) := e^{itx} \cdot f(x), \quad x \in \mathbb{R} \text{ and } f \in L^p(\mathbb{R}, \mu).$$

While the spectrum in each case is $\sigma(M) = i\mathbb{R}$ its fine structure as well as the asymptotic behavior of $(T(t))_{t\geq 0}$ may depend on the measure μ.

To show this we consider weak operator convergence of $T(t)$ to zero, i.e.,

$$\lim_{t\to\infty} < T(t)f, g >= 0$$

for all $f \in L^p$, $g \in L^q$, where q satisfies $\frac{1}{p} + \frac{1}{q} = 1$. It suffices to verify convergence for all exponential functions

$$h_r(x) := e^{irx}, \quad r \in \mathbb{R}.$$

But

$$< T(t)h_r, h_s > = \int_{\mathbb{R}} e^{itx} \cdot e^{i(r+s)x}\, d\mu(x) = \hat{\mu}(t+r+s),$$

where $\hat{\mu}(t) := \int_{\mathbb{R}} e^{itx}\, d\mu(x)$ denotes the Fourier-Stieltjes transform of the measure μ. But $\hat{\mu}$ is a continuous function on $\mathbb{R}$ vanishing at infinity whenever μ is absolutely continuous with respect to the Lebesgue measure. This is the Riemann-Lebesgue lemma (see [Ru2], Thm. 7.5) and will be stated as a convergence result for the above semigroup.

1.2 Riemann-Lebesgue Lemma. *If the probability measure μ on $\mathbb{R}$ is absolutely continuous with respect to Lebesgue measure then the semigroup defined by $T(t)f(x) := e^{itx} \cdot f(x)$ $(x \in \mathbb{R},\ t \geq 0,\ f \in L^p)$ satisfies*

$$\lim_{t \to \infty} T(t) = 0$$

for the weak operator topology on any space $L^p(\mathbb{R}, \mu)$, $1 \leq p < \infty$.

Our next example comes from Ergodic Theory and we look at a measure preserving semiflow

$$\phi_t : \Omega \longrightarrow \Omega, \quad t \in \mathbb{R}, \quad t \geq 0,$$

on some probability space (Ω, Σ, μ). We refer to [Kr] or [C-F-S] and recall that one of the basic properties of such flows is *weak mixing*, defined by

$$\lim_{t \to \infty} \frac{1}{t} \int_0^t |\mu(\varphi_s(M) \cap N) - \mu(M) \cdot \mu(N)|\, ds = 0$$

for all measurable sets M, $N \in \Sigma$.

This property corresponds to a certain asymptotic behavior of the strongly continuous semigroup of linear operators

$$T(t)f(x) := f(\phi_t(x))$$

for $x \in \Omega$ and $f \in L^p(\Omega, \mu)$ for any $1 \leq p < \infty$. In perfect harmony with the philosophy of this paper this behavior of the semigroup $(T(t))_{t\geq 0}$ is characterized by a spectral property of its generator. We collect these standard results from Ergodic Theory (see [C-F-S], Chap. I, §6 & 7) in the so-called

1.3 Weak Mixing Theorem. *Let $(\phi_t)_{t\geq 0}$ be a measure preserving semiflow on a separable probability space (Ω, Σ, μ) and take $(T(t))_{t\geq 0}$ to be the induced linear semigroup on $L^p(\Omega, \mu)$, $1 \leq p < \infty$, with generator $(A, D(A))$. Then the following assertions are equivalent.*

(a) *The flow $(\phi_t)_{t\geq 0}$ is weakly mixing.*

(b) *For the semigroup $(T(t))_{t\geq 0}$ there exists a sequence $t_i \to \infty$ such that*

$$\lim_{i\to\infty} T(t_i) = P$$

for the weak operator topology with P defined by $Pf := \left(\int_\Omega f \, d\mu\right) \cdot \mathbb{1}$, $f \in L^p(\Omega, \mu)$.

(c) *The generator $(A, D(A))$ has 0 as the only eigenvalue and the corresponding eigenspace consists of the constant functions.*

In the final example we look at a similar situation in Topological Dynamics (see [Au], Chap.2). Let $(\phi_t)_{t\geq 0}$ be a continuous semiflow on the compact space X. Among the many properties describing regular or irregular behavior of the flow we recall the one which allows a rather complete description.

The semiflow $(\phi_t)_{t\geq 0}$ is called *equicontinuous* (or *stable*) if $\{\phi_t : t \geq 0\}$ is an equicontinuous set of maps from X into X.

Again we look at the induced semigroup of linear operators on the function space $C(X)$ defined by

$$T(t)f(x) := f(\phi_t(x))$$

for $x \in X$ and $f \in C(X)$. It follows from the Arzela-Ascoli theorem that equicontinuity of the flow corresponds to a compactness property of the induced linear semigroup. We state this in the following

1.4 Theorem. *Let $(\phi_t)_{t\geq 0}$ be a continuous semiflow on the compact space X and denote by $(T(t))_{t\geq 0}$ the induced linear semigroup on $C(X)$. The following properties are equivalent.*

(a) *The semiflow $(\phi_t)_{t\geq 0}$ is equicontinuous.*

(b) *The semigroup $(T(t))_{t\geq 0}$ is relatively compact for the strong operator topology on $\mathcal{L}(C(X))$.*

We remark that the above theorem does not give a characterization based on the spectrum of the generator of $(T(t))_{t\geq 0}$. However, Theorem 1.4 should be compared with the results in Section 4.

After these examples, we now specify our notations and indicate our goals.

Let E be a Banach space and $(T(t))_{t\geq 0}$ be a strongly continuous semigroup of bounded linear operators on E. By $(A, D(A))$ we denote its generator A with domain $D(A)$. In addition, we use the standard spectral theoretic notation $\sigma(A)$, $P\sigma(A)$, $\rho(A)$, $R(\lambda, A)$ for the spectrum, point spectrum, resolvent set and resolvent of A, respectively.

On the space $\mathcal{L}(E)$ of all bounded linear operators on E we consider the uniform topology, the strong operator topology and the weak operator topology. It is our aim to characterize — preferably in terms of the spectrum of A — the (relative) compactness of

$$\{T(t) : t \geq 0\}$$

in $\mathcal{L}(E)$ with respect to uniform, strong and weak operator topology. Of particular importance will be the case that $\lim_{t\to\infty} T(t)$ exists in one of these topologies.

Section 2

Uniform topology: the infinite dimensional Liapunov theorem

In order to generalize the Liapunov Theorem 1.1 to semigroups on Banach spaces it is useful to introduce the *growth bound*

$$\omega := inf\{w \in \mathbb{R} : \exists M \geq 1 \text{ such that } \|T(t)\| \leq M \cdot e^{wt} \ \forall t \geq 0\}$$

and the *spectral bound*

$$s(A) := sup\{Re\,\lambda : \lambda \in \sigma(A)\}$$

of a strongly continuous semigroup $(T(t))_{t\geq 0}$ with generator A. It is clear that the conditions (a), (b) and (c) in Liapunov's Theorem are equivalent if

$$s(A) = \omega\,,$$

which is not true in general. We refer to [Na1], A-III for the counterexamples and a detailed discussion of the situation. Here we only state the main result using the standard terminology.

2.1 Theorem. *Let $(T(t))_{t\geq 0}$ be an eventually norm continuous (e.g., analytic, eventually differentiable or eventually compact) semigroup with generator $(A, D(A))$. Then the following assertions are equivalent.*

(a) $\lim_{t\to\infty} \|T(t)\| = 0$.

(b) *The growth bound satisfies* $\omega < 0$.

(c) *The spectral bound satisfies* $s(A) < 0$.

In Hilbert spaces there is also an analogue of condition 1.1.d which can be found in [G-N].

Section 3

Weak operator topology: the Jacobs-Glicksberg-deLeeuw splitting

We now turn our attention to the weak operator topology. The nice feature of this topology is the fact that a semigroup becomes relatively compact with respect to the weak operator topology under quite weak assumptions. These standard facts from Functional Analysis will be stated first.

3.1 Proposition. *For a semigroup $(T(t))_{t\geq 0}$ on a Banach space E the following assertions are equivalent.*

(a) $\{T(t) : t \geq 0\}$ *is relatively compact in* $\mathcal{L}(E)$ *for the weak operator topology.*

(b) $\{T(t)f : t \geq 0\}$ *is relatively* $\sigma(E, E')$*-compact in* E *for every* $f \in E$.

(c) $\{\|T(t)\| : t \geq 0\}$ *is bounded and* $\{T(t)f : t \geq 0\}$ *is relatively* $\sigma(E, E')$*-compact for every* f *in a dense subset of* E.

If the Banach space E is reflexive and $\{\|T(t)\| : t \geq 0\}$ is bounded, then each oı the above holds. Hence, we have automatic weak operator compactness for bounded semigroups on reflexive Banach spaces. For these an important structure theorem is available yielding a splitting of E and $(T(t))$ into a 'reversible' and an 'irreversible' part. This theorem follows from the general theory of topological semigroups (see e.g. [Kr] or [Ly]) and we state its version for one-parameter semigroups.

3.2 Theorem. *Let* $(T(t))_{t\geq 0}$ *be a semigroup with generator* $(A, D(A))$ *on the Banach space* E *and assume that* $(T(t))_{t\geq 0}$ *is relatively compact for the weak operator topology. Then there exists a splitting*

$$E = E_r \oplus E_o$$

into $(T(t))$*-invariant closed subspaces where*

(i) *the reversible part*

$$E_r := \overline{lin}\{f \in D(A) : Af = i\lambda f \text{ for some } \lambda \in \mathbb{R}\}$$

and

(ii) *the irreversible part*

$$E_o := \{g \in E : 0 \text{ is a } \sigma(E, E')\text{-accumulation point of } \{T(t)g : t \geq 0\}\}.$$

The action of the semigroup $(T(t))_{t\geq 0}$ on the reversible part E_r is determined by the eigenvectors f, i.e., $Af = i\lambda f$ and hence, $T(t)f = e^{i\lambda t}f$ for $t \geq 0$. Thus $(T(t))_{t\geq 0}$ acts isometrically on these eigenvectors. In fact much more can be said.

3.3 Corollary. *Let* $(T(t))_{t\geq 0}$ *be as above and assume* $E = E_r$ *(i.e.,* $(T(t))_{t\geq 0}$ *has 'discrete spectrum'). Then its strong operator closure*

$$\overline{\{T(t) : t \geq 0\}} \subset \mathcal{L}(E)$$

is a compact group of invertible operators.

Such groups are quite well understood. In particular if they consist of positive operators on Banach lattices they can be described explicitly by rotations on solenoidal compact groups (see [Na1]).

The action of $(T(t))_{t\geq 0}$ on the irreversible part is more difficult to understand. The reason is that even the existence of a sequence $(t_i)_{i\in\mathbb{N}}$ in $\mathbb{R}_+$ such that

$$\text{weak-}\lim_{i\to\infty} T(t_i)f = 0$$

does not imply

$$\text{weak-}\lim_{t\to\infty} T(t)f = 0 .$$

Examples for this phenomenon are exhibited by weakly mixing but not strongly mixing flows in Ergodic Theory (see 1.3) and are studied in a more general context in [R-R-S]. Very informative examples can be obtained in the spirit of 1.2 above using interesting facts from Harmonic Analysis. The following is an adaptation from [We] (see also [K-W], Chap. I).

3.4 Example. By [Ru1], Thm. 5.2.2 there exists a closed, uncountable subset X in $\Gamma = \{z \in \mathbb{C} : |z| = 1\}$, called a "Kronecker set", having the property that for some subsequence $(n_i)_{i\in\mathbb{N}}$ in $\mathbb{N}$ one has

$$\lim_{i\to\infty} z^{n_i} = 1 \ \textit{uniformly} \text{ for } z \in X . \qquad (*)$$

Take now a diffuse measure μ supported by X and define the unitary one-parameter group $(T(t))_{t\in\mathbb{R}}$ by

$$T(t)f(z) := z^t \cdot f(z)$$

for $f \in L^2(X,\mu)$ and $z \in X$. Since μ is diffuse the point spectrum of the generator of $(T(t))_{t\in\mathbb{R}}$ is empty. This implies, by Theorem 3.2, that weak-$\lim_{t_i\to\infty} T(t_i)f = 0$ for $f \in L^2(X,\mu)$ and some sequence $(t_i)_{i\in\mathbb{N}}$ in $\mathbb{R}$. On the other hand the above property $(*)$ implies that

$$\lim_{i\to\infty} \|T(n_i) - Id\| = 0 ,$$

and hence, $\lim_{t\to\infty} T(t)$ does not exist.

3.5 Corollary. *For a relatively weakly compact semigroup $(T(t))_{t\geq 0}$ the following assertions are equivalent.*

(a) *The generator A has no purely imaginary eigenvalues, i.e., $P\sigma(A)\cap i\mathbb{R} = \emptyset$.*

(b) *The semigroup $(T(t))_{t\geq 0}$ has 0 as an accumulation point for the weak operator topology.*

Section 4

Strong operator topology: the A-B-L-P theorem and its converse

Convergence and compactness for the strong operator topology is probably the most interesting theoretically and the most useful for applications. We will sketch some of the very recent developments but want to avoid certain technicalities. Therefore, in this section we assume the following.

4.1 Assumptions. Let E be a reflexive Banach space and $(T(t))_{t\geq 0}$ be a bounded, strongly continuous semigroup on E having generator $(A, D(A))$.

From the boundedness of $(T(t))_{t\geq 0}$ it follows that the spectrum $\sigma(A)$ of A is contained in $\{\lambda \in \mathbb{C} : Re\,\lambda \leq 0\}$, i.e., $s(A) \leq 0$. Moreover, since E is reflexive, $\{T(t) : t \geq 0\}$ is always relatively compact for the weak operator topology and E splits into a reversible part E_r and an irreversible part E_o (see Theorem 3.2). On E_r the restricted semigroup becomes relatively compact, even for the strong operator topology, by Corollary 3.3. So, it remains to improve the convergence to 0 on E_o. Since $\lim_{i\to\infty} \|T(t_i)g\| = 0$ for some sequence $t_i \to \infty$ implies $\lim_{t\to\infty} \|T(t)g\| = 0$ we have the following equivalence.

4.2 Lemma. *Under the above assumptions the semigroup $(T(t))_{t\geq 0}$ is relatively compact for the strong operator topology if and only if the irreversible part E_o is given by*

$$E_o = \{g \in E : \lim_{t\to\infty} \|T(t)g\| = 0\}.$$

We call semigroups satisfying this compactness condition *(strongly) stable* and it is our aim to characterize stability of the semigroup through spectral properties of the generator. We will be guided by the following

4.3 Leitmotif. The convergence "$T(t)g \longrightarrow 0$" for $g \in E_o$ is "better" the "thinner" is the *peripheral spectrum*,

$$\sigma(A_o) \cap i\mathbb{R},$$

of the restriction A_o of the generator A to the irreversible part of E_o.

The first step towards our goal and an essential argument in the proofs of the subsequent theorems is based on a classical result of Gelfand in which he extends a well understood phenomena from unitary operators on Hilbert spaces to a class of operators on arbitrary Banach spaces.

4.4 Lemma. *(see [A-R]) Let $T \in \mathcal{L}(E)$ be an invertible operator such that $\{\|T^n\| : n \in \mathbb{Z}\}$ is bounded. Then $T = Id$ if (and only if) $\sigma(T) = \{1\}$.*

Let now $(T(t))_{t\in\mathbb{R}}$ be a bounded, strongly continuous group on E. Then $\sigma(A)$ is contained in $i\mathbb{R}$ and is non-empty. In fact, for bounded groups a 'weak spectral mapping theorem' holds (see [Na1], A-III, Thm. 7.4). Hence, if $\sigma(A) = \{0\}$ we obtain $\sigma(T(t)) = \{1\}$ and the following result. Observe that reflexivity of E is not needed in Lemma 4.4 and Theorem 4.5.

4.5 Theorem. *If $(T(t))_{t\in\mathbb{R}}$ is a bounded, strongly continuous group, then*
$T(t) = Id$, $t \in \mathbb{R}$, if (and only if) $\sigma(A) = \{0\}$.

It is clear that without the boundedness of the group the above equivalence is far from being true (e.g., take $T(t) = \begin{pmatrix} 1 & t \\ 0 & 1 \end{pmatrix}$ with $A = \begin{pmatrix} 0 & 1 \\ 0 & 0 \end{pmatrix}$ on $E = \mathbb{C}^2$).
It was Katznelson-Tzafriri, who in 1985, made an important step forward ([K-T], see also [Ph1]). The semigroup version of their theorem reads as follows (see [Ph2], Cor. 3.4 and [A-Pr], Thm. 3.10).

4.6 Theorem. *Let $(T(t))_{t\geq 0}$ be a bounded semigroup whose generator satisfies $\sigma(A) \cap i\mathbb{R} \subset \{0\}$. Then*

$$\lim_{t\to\infty} \|T(t)(T(s) - Id)R(\lambda, A)\| = 0$$

for all $s > 0$ *and* $\lambda > 0$.

This uniform convergence result implies pointwise convergence of $T(t)$ on all elements of the images of $T(s) - Id$ for all $s > 0$. Since E is reflexive it follows that each $T(s)$ is a mean ergodic operator and E splits into the fixed space of $T(s)$ and the closure of the range of $T(s) - Id$ (see [Kr], Chap. 2, Thm. 1.3). Therefore one obtains the following

4.7 Corollary. *Let the Assumptions 4.1 be satisfied. Then the condition*

$$\sigma(A) \cap i\mathbb{R} \subset \{0\}$$

implies that $\lim_{t\to\infty} T(t)f$ *exists, in norm, for every* $f \in E$. *Moreover,* $P := \lim_{t\to\infty} T(t) \neq 0$ *if and only if* $0 \in P\sigma(A)$.

Shortly after, in 1988, Arendt-Batty [A-B] and independently Lyubich-Phong [L-P] succeeded in obtaining a beautiful and more general spectral condition sufficient for strong convergence and strong compactness.

4.8 A-B-L-P-Theorem. *Let* $(T(t))_{t\geq 0}$ *satisfy Assumption 4.1. If*

$$P\sigma(A) \cap i\mathbb{R} = \emptyset \quad \text{and} \quad \sigma(A) \cap i\mathbb{R} \quad \text{is countable,}$$

then

$$\lim_{t\to\infty} \|T(t)f\| = 0$$

for every $f \in E$.

As we have seen above, the semigroup is always stable on its reversible part. Therefore we obtain stability if the spectrum $\sigma(A_0)$ of the restriction A_0 to the irreversible part has countable intersection with $i\mathbb{R}$.

4.9 Corollary. *Let* $(T(t))_{t\geq 0}$ *satisfy Assumption 4.1 and denote by* A_0 *the restriction of* A *to the irreversible part (see Theorem 3.2). If* $\sigma(A_0) \cap i\mathbb{R}$ *is countable, then* $(T(t))_{t\geq 0}$ *is stable.*

There are many applications and extensions of the Arendt-Batty-Lyubich-Phong-Theorem (see [A-Pr] or [B-P]) but we pursue the discussion of its theoretical significance. The first observation to be made concerns the necessity of the above spectral condition for stability.

4.10 Example. Let $(T(t))_{t\geq 0}$ be the shift semigroup, given by

$$T(t)f(x) := f(x+t)\,, \quad x,t \geq 0\,,$$

for $f \in L^2(\mathbb{R}_+, dx)$. Its generator A has spectrum

$\sigma(A) = \{\lambda \in \mathbb{C} : Re\lambda \leq 0\}$, hence $\sigma(A) \cap i\mathbb{R} = i\mathbb{R}$ is uncountable.

But, the semigroup satisfies

$$\lim_{t\to\infty} \|T(t)f\| = 0\,,$$

for every $f \in L^2(\mathbb{R}_+, dx)$, hence is stable.

The search for a condition equivalent to

'$\sigma(A) \cap i\mathbb{R}$ is countable'

is based on the observation that the spectrum remains unchanged when passing to certain associated operators on new spaces (e.g., on the dual Banach space or on some ultrapower space), but that stability is not preserved in general.

In order to report on the new results in this direction we need the notion of *semigroup-ultrapower* $E_{\mathcal{U}}^{(T(t))}$, for some free ultrafilter $\mathcal{U}$ on $\mathbb{N}$, and corresponding to the strongly continuous semigroup $(T(t))_{t\geq 0}$ on the Banach space E. See [Na1], A-I, 3.6, and [H-R] for precise definitions. There it is shown that $(T(t))_{t\geq 0}$ extends to a strongly continuous semigroup $(T_{\mathcal{U}}(t))_{t\geq 0}$ on $E_{\mathcal{U}}^{(T(t))}$ whose generator $A_{\mathcal{U}}$ satisfies $\sigma(A_{\mathcal{U}}) = \sigma(A)$. Thus, if $E_{\mathcal{U}}$ is still a reflexive Banach space (such spaces are called *superreflexive*, see [N-R]) and $\sigma(A) \cap i\mathbb{R}$ is countable, then it follows from Corollary 4.9 that the ultrapower extension $(T_{\mathcal{U}}(t))_{t\geq 0}$ remains stable. For such a behavior the following terminology seems to be adequate.

4.11 Definiton. *A semigroup $(T(t))_{t\geq 0}$ is called superstable if each ultrapower extension $(T_{\mathcal{U}}(t))_{t\geq 0}$ is stable, for every ultrafilter $\mathcal{U}$ on $\mathbb{N}$.*

From the above discussion it follows that, on superreflexive Banach spaces, a bounded semigroup with countable peripheral spectrum $\sigma(A)\cap i\mathbb{R}$ is superstable. That the converse also holds has recently been shown by [H-R], based on previous results by [N-R] for discrete semigroups $(T^n)_{n\in\mathbb{N}}$.

4.12 Theorem. *For a bounded semigroup* $(T(t))_{t\geq 0}$ *with generator* $(A, D(A))$ *on a superreflexive Banach space* E *the following assertions are equivalent.*

(a) *The peripheral spectrum* $\sigma(A)\cap i\mathbb{R}$ *is countable.*

(b) *The semigroup* $(T(t))_{t\geq 0}$ *is superstable.*

We conclude this survey paper by mentioning some open problems.

Section 5

Open problems

5.1 Weak convergence. No spectral conditions on the generator A characterizing, or related to, the property

$$\text{“weak-}\lim_{t\to 0} T(t)f = 0 \quad \text{for every} \quad f \in E\text{”}$$

seem to be known. On the other hand, for unitary groups on Hilbert spaces, we can apply the spectral theorem and are therefore in a situation like in 1.2. See also Chap.1, §7, Theorem 3 in [C-F-S].

5.2 Strong convergence. While

$$\text{“}\lim_{t\to 0} T(t)f = 0\,, \quad \text{in norm, for every} \quad f \in E\text{”}$$

is implied by the spectral condition in Theorem 4.8, a spectral characterization of this convergence is still unknown.

5.3 Trivial spectrum implies trivial operator. Such an implication can only hold under precise additional hypotheses as, e.g., in Theorem 4.5. Therefore it is interesing to find other situations allowing similar results. For example, we think of the following:

Let $(T(t))_{t\geq 0}$ be a strongly continuous semigroup of positive Markov operators on a Banach space $C(X)$, with X compact. Is it true that $\sigma(A) = \{0\}$ implies $A = 0$ and hence, $T(t) = Id$ for all $t \geq 0$?

For positive groups such results are obtained in [A-G] and [Gr].

[Au] Auslander, J.: "Minimal Flows and their Extensions", North Holland, 1988.

[A-B] Arendt, W., Batty, C.J.K.: *Tauberian theorems and stability of one-parameter semigroups,* Trans. Amer. Math. Soc. **306** (1988), 837-852.

[A-G] Arendt, W., Greiner, G.: *A spectral mapping theorem for one-parameter groups of positive operators in $C_0(X)$,* Semigroup Forum **30** (1984), 297-330.

[A-Pr] Arendt, W., Prüss, J.: *Vector-valued Tauberian theorems and asympototic*
behavior of linear Volterra equations, Siam J. Math. Anal. **23** (1992), 412-448.

[A-R] Allan, G.R., Ransford, T.J.: *Power- dominated elements in a Banach algebra* **94** (1989), 63-79.

[B-P] Batty, C.J.K., Phong, V.Q.: *Stability of individual elements under one-parameter semigroups,* Trans. Amer. Math. Soc. **322** (1990), 805-818.

[C-F-S] Cornfeld, J.P., Fomin, S.V., Sinai, Ya.G.: "Ergodic Theory", Springer-Verlag, 1982.

[Gr] Greiner, G.: *A spectral decomposition of strongly continuous groups of positive operators,* Quart. J. Math. Oxford **35** (1984), 37-47.

[G-N] Groh, U., Neubrander, F.: *Stabilität starkstetiger Operatorhalbgruppen auf*
C^*- Algebren, Math. Ann. **256** (1981), 509-516.

[H-J] Horn, R., Johnson, C.R.: "Topics in Matrix Analysis", Cambridge Univ. Press, 1991.

[H-R] Huang, S., Räbiger, F.: *Superstable C_0-semigroups on Banach spaces,* Preprint (1992).

[Kr] Krengel, U.: "Ergodic Theorems", de Gruyter, 1985.

[K-T] Katznelson, Y., Tzafriri, L.: *On power bounded operators,* J. Funct. Anal. **68** (1986), 313-328.

[K-W] Kaashock, M.A., West, T.T.: *Locally Compact Semialgebras,* Mathematics Studies **9**, North-Holland 1974.

[Ly] Lyubich, Yu.I.: "Introduction to the Theory of Banach Representations of Groups", Birkhäuser Verlag, 1988.

[L-P] Lyubich, Yu.I., Phong, V.Q.: *Asymptotic stability of linear differential equations in Banach spaces,* Studia Math. **88** (1988), 37-42.

[Na1] Nagel, R.(ed.): "One-parameter Semigroups of Positive Operators", Lect. Notes Math. **1184**, Springer-Verlag, 1986.

[Na2] Nagel, R.: *On the linear operator approach to dynamical systems,* Conf. Sem. Mat. Univ. Bari.

[N-R] Nagel, R., Räbiger, F.: *Superstable operators in Banach spaces,* Israel J. Math. (to appear).

[Ph1] Phong, V.Q.: *A short proof of the Y.Katznelson's and L.Tzafriri's theorem,* Proc. Amer. Math. Soc. Preprint (1992).

[Ph2] Phong, V.Q.: *Theorems of Katznelson-Tzfariri type for semigroups of operators,* J. Funct. Anal. **103** (1992), 74-84.

[Ru1] Rudin, W.: "Fourier Analysis on Groups", Interscience Publ. 1967.

[Ru2] Rudin, W.: "Functional Analysis", McGraw-Hill Book Comp., 1973.

[R-R-S] Rosenblatt, J., Ruess, W.M., Sentilles, D.: *On the critical part of a weakly almost periodic function,* Houston J. Math. **17** (1991), 237-249.

[We] West, T.T.: *Weakly compact monothetic semigroups of operators in Banach spaces,* Proc. Royal Irish Acad. **67**, Sect. A (1968), 27-37.

Rainer Nagel
Universität Tübingen
Mathematisches Institut
Auf der Morgenstelle 10
7400 Tübingen
Germany
Tel.: 0044-7071-293242
e-mail: nagel@michelangelo.zdv.uni-tuebingen.de

Continuation for Quasiholomorphic Semigroups

J. W. Neuberger

Suppose X is a Banach space and T is a strongly continuous semigroup of members of $L(X,X)$. It has been known for a long time that asymptototic properties of

$$|T(t) - I|$$

as $t \to 0$ relate closely to smoothness properties of T.

In particular one has that,

$$\lim_{t\to 0+} |T(t) - I| = 0 \tag{1}$$

if and only if T has a generator which is in $L(X,X)$ (cf [4], p 282). Not so well known is the fact that if

$$\limsup_{t\to 0+} |T(t) - I| < 2 \tag{2}$$

then T is holomorphic ([1],[2, p 317],[6] [9]). In particular if (2) holds then

$$AT(t) \in L(X,X) \text{ for all t} > 0 \tag{3}$$

where A is the generator of T defined by

$$D(A) = \{x \in X : \lim_{t\to 0+} (1/t)(T(t)x - x) \text{ exists}\}$$

and, if $x \in D(A)$,

$$Ax = \lim_{t\to 0+} (1/t)(T(t)x - x).$$

Consequently, if

$$\limsup_{t\to 0+} |T(t) - I| < 1, \tag{4}$$

G. R. Goldstein and J. A. Goldstein (eds.),
Semigroups of Linear and Nonlinear Operations and Applications, 241–249.

then (2) holds and, in addition,

$$T(t)^{-1} \in L(X, X)$$

for some $t > 0$. Hence if (4) holds, the generator A of T must be bounded and so (4) implies (1).

This note concerns implications of the condition

$$\liminf_{t \to 0+} |T(t) - I| < 2. \tag{5}$$

We will call semigroups satisfying (5) quasiholomorphic. Examples will show that (5) does not imply (2) and that (5) itself does not always hold.

Denote by *fot*(T) (functional of trajectory of T) the set of all functions g on $[0, \infty)$ so that

$$g(t) = f(T(t)x), \ t \geq 0.$$

It is known that ([5], [6]) if T satisfies (5), then *fot*(T) is quasianalytic in the sense that no two members of *fot*(T) agree on an open subset of $[0, \infty)$. If *fot*(T) is quasianalytic, we say that T itself is quasianalytic. Thus the above references give that every quasiholomorphic semigroup is quasianalytic. Observe that (2) implies that each member of *fot*(T) is real-analytic on $[0, \infty)$. The new result of this note gives that, under (5), members of *fot*(T) have a continuation property in the sense that each $g \in$ *fot*(T) can be computed from its restriction to any subinterval of $[0, \infty)$ (as surely holds under (2)). The ability to do this extension, assuming (5), is a consequence of recent results in [8].

Some Examples. Suppose that $q = n_1, n_2, \ldots$ is an increasing sequence of positive integers and

$$T_q(t)x = (x_1 \exp(in_1 t), \ x_2 \exp(in_2 t), \ldots), \ t \geq 0, \tag{6}$$

for all $x = (x_1, x_2, \ldots) \in l_2$. Then T_q is a strongly continuous semigroup on l_2 (actually extendable to a group). The following calculates $|T_q(t) - I|$:

Lemma. Suppose (6) holds. Then

$$|T_q(t) - I| = 2 \cdot \sup_{k=1,2,\dots} |\sin(n_k t/2)|, \; t \geq 0 \tag{7}$$

Proof of Lemma. Suppose $t \geq 0$. Then

$$\begin{aligned} |T_q(t) - I|^2 &= \sup_{x \in l_2, \|x\|=1} \|T(t)x - x\|^2 \\ &= \sup_{x \in l_2, \|x\|=1} \sum_{k=1}^{\infty} |x_k(\exp(in_k t) - 1|^2 \\ &\geq \sup_{k=1,2,\dots} |\exp(in_k t) - 1|^2 \\ &= 4 \cdot \sup_{k=1,2,\dots} |\sin(n_k t/2)|^2 \end{aligned}$$

so that

$$|T_q(t) - I| \geq 2 \cdot \sup_{k=1,2,\dots} |\sin(n_k t/2)|.$$

But also,

$$\sup_{x \in l_2, \|x\|=1} \sum_{k=1}^{\infty} |x_k(\exp(in_k t) - 1)|^2 \leq 4\|x\|^2 \cdot \sup_{k=1,2,\dots} (\sin(n_k t/2))^2$$

so that

$$|T_q(t) - I| \leq 2 \cdot \sup_{k=1,2,\dots} |\sin(n_k t/2)|.$$

Therefore

$$|T_q(t) - I| = 2 \cdot \sup_{k=1,2,\dots} |\sin(n_k t/2)|,$$

and the argument is complete.

Example 1. Take $w = 1, 2, 3, \dots$. Then since

$$\sup_{k=1,2,\dots} |\sin(k\, t/2)| = 1$$

it follows that

$$|T_w(t) - I| = 2, \quad t \geq 0.$$

Example 2. Take $v = 3^1, 3^2, \ldots,$ $t_m = 2\pi 3^{-m}$, $m = 1, 2, \ldots$. Then if $m \in Z^+$,

$$\sup_{k=1,2,\ldots} |\sin(3^k \cdot 2\pi \cdot 3^{-m})/2| = \sin(\pi/3) = 3^{1/2}/2$$

so that

$$|T_v(t_m) - I| \leq 3^{1/2}, \quad m = 1, 2, \ldots$$

and hence

$$\lim_{t \to 0+} \inf |T_v(t) - I| \leq 3^{1/2}.$$

Note, however, that

$$\lim_{t \to 0+} \sup |T_v(t) - I| = 2.$$

It is clear that T_v is not holomorphic since if it were, then $AT_v(t)$ would be bounded for all $t > 0$, a contradiction, since A is not bounded and $T_v(t)^{-1} \in L(l_2, l_2)$ for all $t \geq 0$.

The following example shows that *fot*(T) need not be a quasianalytic collection for T a strongly continuous semigroup:

Example 3. Denote by X the Banach space (with sup norm) of all bounded uniformly continuous functions h from R to C so that $h(x) = 0$, $x \leq 0$. Define the semigroup T on X by

$$(T(t)h)(x) = h(x - t), \quad h \in X,\ t \geq 0,\ x \in R. \tag{8}$$

Now take $q \in X$ so that

$$q(x) = 0,\ x \leq 0,\ q(x) = x(1 - x),\ 0 \leq x \leq 1,\ q(x) = 0,\ x \geq 1,$$

and $f \in X^*$ so that $f(h) = h(1/2)$, $h \in X$. Define g on $[0, \infty)$ so that

$$g(t) = f(T(t)q) = h(1/2 - t), \quad t \geq 0.$$

Then $g(0) \neq 0$, but $g(t) = 0$, $t \geq 1/2$. Since the zero function on $[0, \infty)$ is also in *fot*(T), it follows that *fot*(T) is not quasianalytic.

In summary: Not all strongly continuous semigroups are quasianalytic and there are quasianalytic semigroups which are not holomorphic.

The following is needed in order to express our main result:

Suppose a, b, $c \in R$, b is in the open interval (a, c), $\epsilon \in (0, 1)$, $M > 0$, $\delta \in R$, $\delta \neq 0$, $|\delta| < \min(|b-a|, |c-b|)$ and g is a continuous function whose domain includes $[a, c]$, We pose:

Problem $P(a,\, b,\, c,\, \epsilon,\, M,\, \delta,\, g)$: Find $a_n, \ldots, a_m$ so that

$$|\sum_{k=0}^{p} \binom{p}{k} (-1)^{p-k} a_k| \leq (2-\epsilon)^p \quad \textit{and} \quad |a_p| \leq M, \;\; p = n, \ldots, m, \tag{9}$$

where

$$a_p = g(a + \delta p),\; p = 0, 1, \ldots, n-1, \quad n = [|(b-a)/\delta|] + 1, \;\; m = [|(c-a)/\delta|].$$

We use the following notation: If $r = a_n, \ldots, a_m$ is a solution to problem $P(a,\, b,\, c,\, \epsilon,\, M,\, \delta,\, g)$ and q is the set of parameters a, b, c, ϵ, M, δ, g, then $g_{q,r}$ is the function on $[a, c]$ defined as follows:

(i) $g_{q,r}(t) = g(t)$, $t \in [a, b]$,

(ii) $g_{q,r}(a + \delta p) = a_p$, $p = n, \ldots, m$,

(iii) $g_{q,r}$ is continuous and linear on each of the intervals $[b, a + n\delta]$, $[a + p\delta, a + (p+1)\delta]$, $p = n, \ldots, m-1$, and

(iv) $g_{q,r}$ is constant on $[a + m\delta,\, c]$.

Theorem. Suppose T is a strongly continuous nonexpansive semigroup on the Banach space X, $\epsilon > 0$, and $\delta_1, \delta_2, \ldots$ is a decreasing sequence of positive numbers converging to zero. Suppose also that

$$\limsup_{k \to \infty} |T(\delta_k) - I| \leq 2 - \epsilon_0.$$

Denote by f a member of X^* and by x a member of X so that $|f| = 1 = \|x\|$. Define g on $[0, \infty)$ so that $g(t) = f(T(t)x)$, $t \geq 0$. Finally, suppose that a, b, c are three nonnegative numbers so that b is in (a, c) and $0 < \epsilon < \epsilon_0$. There is a $K \in Z^+$ so that if $k \in Z^+$, $k \geq K$, then problem $P(a, b, c, \epsilon, M, \delta_k, g)$ has a solution q_k. Moreover, if for each $k \in Z^+$, $k \geq K$, r_k denotes the sequence $\{g(a + \delta_k v)\}_{v=0}^{w}$ (where w denotes the largest integer such that $(a + \delta_k w) \in [a, b]$), then

$$\{g_{q_k, r_k}\}_{k=K}^{\infty}$$

converges uniformly to g on some interval $[a, d]$ for some $d \in (b, c)$.

The point of the theorem is that the problems:

$$P(a, b, c, \epsilon, M, \delta_k, g), \ k \geq K$$

use only the restriction of g to $[a, b]$ in order to construct functions on $[a, c]$ which agree with g on $[a, b]$. Such a sequence (it is not unique) converges to g on a subinterval of $[a, c]$ which is larger than $[a, b]$. This is our means of constructively determining g outside of $[a, b]$ from the restriction of g to $[a, b]$. Repitition of this process yields a construction of g on all of $(0, \infty)$. This is our substitute for analytic continuation in the quasiholomorphic case arising from (5).

Proof of Theorem. Suppose $f \in X^*$, $x \in X$, $|f| = 1 = \|x\|$ and $g(t) = f(T(t)x)$, $t \geq 0$. Note that if $n \in Z^+$, $\delta \geq 0$, $u \geq 0$, then

$$\begin{aligned}\Delta_g(n; u, \delta) &= \sum_{k=0}^{n} \binom{n}{k} (-1)^{n-k} g(u + k\delta) \\ &= f[(\sum_{k=0}^{n} \binom{n}{k} (-1)^{n-k} T(\delta)^k) T(u)x] \\ &= f[(T(\delta) - I)^n T(u)x]\end{aligned}$$

and so

$$|\Delta_g(n; u, \delta)| \leq |T(\delta) - I|^n. \tag{10}$$

By hypothesis,

$$\limsup_{k\to\infty} |T(\delta_k) - I| \leq 2 - \epsilon_0.$$

Denote by ϵ a member of $(0, \epsilon_0)$ and denote by K_0 a member of Z^+ so that if $k > K_0$ then $|T(\delta_k) - I| \leq 2 - \epsilon$. Now suppose $k > K_0$, $f \in X^*$, $x \in X$, $|f| = 1 = \|x\|$ and $g(t) = f(T(t)x)$, $t \geq 0$. By (10),

$$|\Delta_g(n; u, \delta_k)| \leq |T(\delta_k) - I|^n.$$

Hence

$$|\Delta_g(n; u, \delta_k)| \leq (2 - \epsilon)^n \text{ if } u \geq 0,\ n \in Z^+. \tag{11}$$

But this is precisely a condition which implies the hypothesis of the main theorem of [8]. This hypothesis in turn implies the conclusion to the present theorem.

Note that the well-known reduction of the study of a general strongly continuous semigroup to the study of a nonexpansive semigroup yields a similar result for general strongly continuous linear semigroups.

In [8], [7], a function is said to chaotic at a point x in the interior of $D(f)$ provided that

$$\limsup_{\delta\to 0+} \phi(S, f, \delta) = 2$$

for all subintervals S of $D(f)$ so that x is in the interior of S where

$$\phi(S, f, \delta) = \sup\{|\Delta_f(n; u, \delta)|^{1/n} : [u, u + n\delta] \subset S\}.$$

In [8] it is shown that all nonchaotic functions f on an open interval have the above mentioned extension property. Using this terminology, what we have demonstrated above is that if T satisfies (5), then all members of *fot*(T) are nowhere chaotic on $(0, \infty)$.

What might be of particular value is a generalization of the above to some classes of semigroups of nonlinear transformations. This problem was first mentioned in [6] but no progress has been made in this direction. What gives this problem particular interest is that some Navier-Stokes systems generate nonlinear semigroups (cf [3]). It is conjectured that the time of onset to turbulence for a given trajectory of such a semigroup might occur at the first time at which a functional of that trajectory becomes chaotic in the above sense. Thus a condition for nonlinear semigroups which replaces (5) might yield a condition which precludes turbulence for some trajectories of a Navier-Stokes semigroup.

References

[1] A. Beurling. On Analytic extension of Semigroups of Operators. *J. Functional Analysis 6.* 387-401 (1970).

[2] A. Beurling. *Collected Works Vol. 2.* Birkhauser 317-331 (1989).

[3] P. Constantine and C. Foias. *Navier-Stokes Equations.* Chicago Lec. Math. (1988).

[4] R. Phillips and E. Hille, *Functional Analysis and Semigroups.* American Math. Soc. Colloq. Pubs. XXXI (1957).

[5] D. G. Kendall. Some Recent Developments in the Theory of Denumerable Markov Processes. Trans Fourth Prague Conference on Information Theory, Statistical Decision Functions, Random Processes. Academia Prague. 11-27 (1967).

[6] J. Neuberger. Quasi-analyticity and Semigroups. Bull. Amer. Math. Soc. 78. 909-922 (1972).

[7] J. Neuberger. Chaos and Higher Order Differences. Proc. Amer. Math. Soc. 101. 45-50 (1987).

[8] J. Neuberger. Predictability in Absence of Chaos. J. Math. Anal. Appl., to appear.

[9] J. Neuberger. Beurling's Analyticity Theorem. Math. Intelligencer, to appear.

Department of Mathematics
University of North Texas
Denton, Texas 76203
USA

Global existence for a reaction-diffusion system with a balance law

Michel PIERRE and Didier SCHMITT

This paper is a contribution to the study of global existence of solutions to reaction-diffusion systems which have the two main following properties :

- first the nonnegativity of the various components of the solution is preserved with time,
- then the total mass of the components is preserved (or more generally non increasing).

Many results have already been obtained for these systems (see [1],[2], [5], [6], [7]). However much needs to be understood yet. For instance, if we restrict ourselves to 2×2 systems, global existence is proved only for systems of the above class for which one of the components is a priori uniformly bounded.

Here, we consider a 2×2 system for which no a priori knowledge of this sort is easily available, namely

$$(1) \qquad u_t - d_1 u_{xx} = -c(x)u^\alpha v^\beta \text{ on } (0,\infty) \times (-1,1)$$

$$(2) \qquad v_t - d_2 v_{xx} = c(x)u^\alpha v^\beta \text{ on } (0,\infty) \times (-1,1)$$

$$(3) \qquad u(t,x) = v(t,x) = 0 \text{ on } (0,\infty) \times \{-1,1\}$$

$$(4) \qquad u(0,x) = u_0(x), v(0,x) = v_0(x) \text{ on } (-1,1)$$

where d_1, d_2 are positive constants, $\alpha, \beta > 1$

$$(5) \qquad u_0, v_0 \in L^\infty(-1,1), u_0, v_0 \geq 0 \text{ on } (-1,1)$$

$$(6) \qquad c \in L^\infty(-1,1).$$

It is classical that the system (1) - (4) has local classical nonnegative solutions on some interval $(0,T)$. Moreover, the maximal time of existence $T_{\max}$ is characterized by the fact that

$$(7) \qquad \begin{cases} (\|u(t)\|_\infty + \|v(t)\|_\infty \leq C, \text{ for all } t \text{ in } (0,T_{\max})) \\ \Longrightarrow (T_{\max} = \infty), \end{cases}$$

G. R. Goldstein and J. A. Goldstein (eds.),
Semigroups of Linear and Nonlinear Operations and Applications, 251–258.

where $||.||_\infty$ denotes the L^∞-norm on $(-1,1)$.

If $d_1 = d_2$, $u+v$ satisfies the linear heat equation so that u and v are uniformly bounded since, by the maximum principle

$$||u(t)+v(t)||_\infty \leq ||u_0+v_0||_\infty \text{ and } u,v \geq 0.$$

By (7), $T_{\max} = \infty$.

When $d_1 \neq d_2$, the problem of global existence becomes non-trivial. It is proved in [2] that, if one of the solutions is a priori bounded on any interval $(0,T)$, then so is the other one and global existence holds. It is the case for instance if $c \geq 0$ on $(-1,1)$, in which case, by maximum principle

$$||u(t)||_\infty \leq ||u_0||_\infty.$$

These systems, with a priori knowledge on one of the components, have been studied in [2], [5], [7].

Here we want to look at the case when the sign of c is not constant on $(-1,1)$ so that none of the solutions is a priori known to be bounded. In order to emphasize the effect of a change of sign, we will more precisely assume that

$$(8) \qquad a.e.x \in (0,1), c(x) \geq 0, a.e.x \in (-1,0), c(x) \leq 0.$$

As first noticed in [3], one can obtain "local" uniform estimates on u and v of the following kind

$$(9) \qquad \begin{cases} \text{for all } \epsilon \text{ in } (0,1), \text{ for all } T \leq T_{\max}, \text{ for all } t \text{ in } (0,T), \\ \text{for all } x \text{ in } (-1,-\epsilon) \cup (\epsilon,1) \\ u(t,x)+v(t,x) \leq C(T,\epsilon,u_0,v_0). \end{cases}$$

In other word, u and v can only blow up at $x=0$ for some t^*. If so, they have to both blow up at the same time since, by the above recalled results, if one of u or v is bounded, so is the other one.

We do not know yet what happens in general in the situation (8). However, we can prove that, if c vanishes fast enough at $x=0$, then global existence holds for (1)-(4). More precisely.

Theorem. *Assume (5), (6), (8) hold and*

$$\text{(10)} \qquad \begin{cases} \textit{There exist } \gamma > \alpha + \beta - 2 \textit{ and } K > 0, \textit{ such that} \\ \textit{for all } x \text{ in } (0,1), \ c(x) \leq K\, x^{\gamma}. \end{cases}$$

Then, the problem (1)-(4) has a global (classical) solution on $(0,\infty) \times (-1,1)$.

Remarks. Note that condition (10) is unilateral since it only involves the behavior of c for $x > 0$. Obviously, by symmetry the same conclusion would be obtained under the assumption

$$\text{(11)} \qquad \begin{cases} \text{There exist } \gamma > \alpha + \beta - 2, \text{ and } K > 0 \text{ such that} \\ \text{for all } x \text{ in } (-1,0), \ |c(x)| \leq K|x|^{\gamma}. \end{cases}$$

We do not know whether the condition $\gamma > \alpha + \beta - 2$ is optimal. In particular it would be interesting to know whether global existence holds with $c(x) = \text{sign}\,(\text{x})$. Numerical computations made with $d_1 = 1$, $d_2 = 0{,}01$, $\alpha = \beta = 5$ show that the non-linear term $u^{\alpha}v^{\beta}$ can have a very high peak around $x = 0$. However, this "peak" seems to disappear after a while.

The proof of the theorem relies on the following lemma together with previous techniques. We set $\omega_T = (0,T) \times (0,1)$.

Lemma 1. *Assume (5), (6), (8) hold. Then, for all* $T \leq T_{\max}$ *and all* $1 < p < \infty$, *there exists* $C = C(T,p)$ *such that*

$$\text{(12)} \qquad \begin{cases} ||xu||_{L^p(\omega_T)} + ||xv||_{L^p(\omega_T)} & \leq C(T)\Big(||u_0 + v_0||_{L^p(0,1)} \\ & \quad + ||u+v||_{L^\infty(0,T;L^1(0,1))}\Big). \end{cases}$$

Proof. We argue by duality introducing, for $\theta \in C_0^\infty(\omega_T)$, $\theta \geq 0$, the unique solution of the following system :

$$\text{(13)} \qquad -(\psi_t + d\psi_{xx}) = \theta \text{ on } \omega_T$$

$$\text{(14)} \qquad \psi(t,0) = \psi(t,1) = 0 \text{ on } (0,T)$$

$$\text{(15)} \qquad \psi(T,x) = 0 \text{ on } (0,1).$$

Recall that ψ exists, is nonnegative and satisfies for all $1 < q < \infty$ (see [4])

(16)
$$||\psi_t||_{L^q(\omega_T)} + ||\psi_{xx}||_{L^q(\omega_T)} + \sup_{t\in(0,T)} ||\psi(t,.)||_{L^q(0,1)} \le C||\theta||_{L^q(\omega_T)}.$$

By Sobolev's imbedding theorem, since for all $v \in W^{2,q}(0,1)$ we have

$$||v_x||_{L^\infty(0,1)} \le C\big(||v_{xx}||_{L^q(0,1)} + ||v||_{L^q(0,1)}\big), \tag{17}$$

from (16) we also deduce

$$||\psi_x||_{L^q(0,T;L^\infty(0,1))} \le C||\theta||_{L^q(\omega_T)}. \tag{18}$$

We first use $d = d_1$ in (13). By (1), (8), we have

$$u_t - d_1 u_{xx} \le 0 \text{ on } \omega_T. \tag{19}$$

We multiply (19) by $x\psi$ which is nonnegative to obtain

$$x\psi u_t - d_1 x\psi u_{xx} \le 0 \text{ on } \omega_T. \tag{20}$$

Integration by parts on $(0,1)$ gives

$$\int_0^1 x\psi u_{xx} = \big[x\psi u_x - u(\psi + x\psi_x)\big]_0^1 + \int_0^1 u(2\psi_x + x\psi_{xx}). \tag{21}$$

Using boundary conditions (14), (3) and the regularity of u, ψ, the integrated part is equal to zero. Therefore, by (20), (21)

$$\int_0^T\int_0^1 x\,\psi u_t \le \int_0^T\int_0^1 d_1 u(2\psi_x + x\psi_{xx}). \tag{22}$$

We integrate by parts in time to obtain

$$-\int_0^T\int_0^1 x\,u(\psi_t + d_1\psi_{xx}) \le \int_0^1 x\psi(0)u_0 + \int_0^T\int_0^1 2d_1 u\,\psi_x \tag{23}$$

or, going back to (13) (with $d = d_1$), with $p = q'$,

$$\int_0^T \int_0^1 x\,u\,\theta \le ||\psi(0)||_{L^q(0,1)} \cdot ||u_0||_{L^p(0,1)} + \tag{24}$$
$$2d_1 ||u||_{L^\infty(0,T;L^1(0,1))} ||\psi_x||_{L^1(0,T;L^\infty(0,1))}.$$

Using (16), (18), the estimate (24) implies by duality that

$$||xu||_{L^p(\omega_T)} \le C(||u_0||_{L^p(0,1)} + ||u||_{L^\infty(0,T;L^1(0,1))}). \tag{25}$$

This gives the first part of the estimate (12).

To estimate xv, we use (13) with $d = d_2$ and

$$v_t - d_2\, v_{xx} = -(u_t - d_1\, u_{xx}). \tag{26}$$

Again, we multiply by $x\psi$, integrate by parts in space and time and we get

$$\Big[\int_0^1 x\psi v\Big]_0^T - \int_0^T \int_0^1 x\,v\psi_t + d_2 v(x\psi_{xx} + 2\psi_x)$$
$$+ \big[- d_2 x\,\psi v_x + d_2 v(\psi + x\psi_x) - d_1 x\psi u_x$$
$$+ d_1 u(\psi + x\psi_x)\big]_0^1$$
$$= -\Big[\int_0^1 x\psi u\Big]_0^T + \int_0^T \int_0^1 x\,u\psi_t + d_1 u(x\psi_{xx} + 2\psi_x).$$

Integrated terms are equal to zero. Using the definition of ψ, we have

$$\int_0^T \int_0^1 x\,v\theta \le \int_0^1 x\psi(0)(v_0 + u_0)$$
$$+ \int_0^T \int_0^1 2\psi_x(d_2 v + d_1 u) + xu(\psi_t + d_1\psi_{xx})$$
$$\le ||\psi(0)||_{L^q(0,1)} ||v_0 + u_0||_{L^p(0,1)}$$
$$+ C||\psi_x||_{L^1(0,T;L^\infty(0,1))} \cdot ||v + u||_{L^\infty(0,T;L^1(0,1))}$$
$$+ ||xu||_{L^p(\omega_T)} ||\psi_t + d_1\psi_{xx}||_{L^q(\omega_T)}.$$

We now apply estimates (16), (18), (25) to deduce by duality

$$(27)\quad ||xv||_{L^p(\omega_T)} \leq C\big(||u_0+v_0||_{L^p(0,1)} + ||v+u||_{L^\infty(0,T;L^1(0,1))}\big)$$

which, together with (25) yields (12).

Lemma 2. *For all $T < T_{\max}$*

$$(28)\qquad ||u+v||_{L^\infty(0,T;L^1(0,1))} \leq ||u_0+v_0||_{L^1(-1,1)}.$$

Proof. We add (1) and (2), integrate in space on $(-1,1)$ and in time using the boundary conditions (3).

Lemma 3. *(see [4]). Assume w is a classical nonnegative solution of*

$$(29)\qquad w_t - d\,w_{xx} \leq a\,w \text{ on } Q_T = (0,T)\times(-1,1)$$

$$(30)\qquad w = 0 \text{ on } (0,T)\times\{-1,1\}$$

$$(31)\qquad w(0,.) = w_0 \in L^\infty(-1,1), w_0 \geq 0$$

where a is a nonnegative function of $L^r(0,T;L^q(-1,1))$ with

$$(32)\qquad \begin{cases} q\in[1,\infty), r\in[\frac{1}{1-\kappa},\frac{2}{1-2\kappa}], & 0<\kappa<1/2 \\ \frac{1}{r}+\frac{1}{2q} = 1-\kappa. \end{cases}$$

Then, there exists C depending on $||a||_{L^r(0,T;L^q(-1,1))}, r, q, T, \kappa$ and $||w_0||_\infty$ such that

$$(33)\qquad ||w||_{L^\infty(Q_T)} \leq C.$$

Proof of the theorem. We apply Lemma 3 with $w = v$, $d = d_2$, $a(t,x) = 1_{[x>0]}Kx^\gamma u^\alpha v^{\beta-1}$. Since $c(x)\leq 0$ on $(-1,0)$ and $c(x)\leq Kx^\gamma$ on (0,1), by (2) we do have

$$(34)\qquad v_t - d_2 v_{xx} \leq a\,v.$$

Now, in Lemma 3, we also choose $q = 1$ and $r = 1/(1-\kappa-\frac{1}{2}) = 1/(1-\sigma)$ where σ is chosen in $(1/2,1)$ so that (see assumption (10))

$$(35)\qquad \delta = \gamma-(\alpha+\beta-1) > -\sigma > -1.$$

We compute the $L^r(0,T;L^1(-1,1))$-norm of a :

$$\int_0^t \Big[\int_0^1 a(t,x)dx\Big]^r dt \leq K^r \int_0^T dt\Big[\int_0^1 x^\delta (xu)^\alpha (xv)^{\beta-1} dx\Big]^r$$
$$\leq K^r \int_0^T dt\Big(\int_0^1 x^{\delta/\sigma}\Big)^{\sigma r} \int_0^1 (xu)^{\frac{\alpha}{1-\sigma}} (xv)^{\frac{\beta-1}{1-\sigma}}$$
$$(r(1-\sigma)=1).$$

Because of (35), $\int_0^1 x^{\delta/\sigma} < \infty$. We again apply Hölder's inequality to obtain

$$(36) \qquad \int_0^T \Big[\int_0^1 a(t,x)dx\Big]^r dt \leq C||xu||^{\alpha/(1-\sigma)}_{L^p(\omega_T)} ||xv||^{(\beta-1)/(1-\sigma)}_{L^q(\omega_T)}$$

for some $p, q < \infty$.

From (34), (36), Lemmas 1, 2 and 3, we obtain that

$$(37) \qquad ||v||_{L^\infty(Q_T)} \leq C(T)$$

where $C(T)$ depends on T and the data.

To finish the proof, we use the usual arguments (see [2], [5]) saying that (26) implies that for all $1 < p < \infty$

$$(38) \qquad ||u||_{L^p(Q_T)} \leq C(1 + ||v||_{L^p(Q_T)}).$$

From (1), (37), (38), we deduce that $||u||_{L^\infty(Q_T)} \leq C(T)$ for all T, whence the global existence.

References

[1] A. Haraux, A. Youkana, *On a Result of K. Masuda Concerning Reaction-Diffusion Equations,* Tohoku Math. J. 40 : 159-163 (1988).

[2] S. Hollis, R.H. Martin, M. Pierre, *Global Existence and Boundedness in Reaction Diffusion Systems,* SIAM J. Math. Anal. 18 : 744-761 (1987).

[3] S. Hollis, J. Morgan, *Interior Estimates for a Class of Reaction Diffusion Systems from L^1 a priori Estimates,* JDE 98 : 260-276 (1992).

[4] O.A. Ladyzenskaja, V.A. Solonnikov, N.N. Uralceva, *Linear and Quasilinear Equations of Parabolic Type,* Trans. Math. Monographs vol. 23 American Mathematical Society, Providence, R.I. 1968.

[5] R.H. Martin, M. Pierre, *Nonlinear Reaction Diffusion Systems,* Nonlinear Equations in the Applied Sciences, ed. W.F. Ames and C. Rogers, Academic Press (1991), Notes and Reports in Mathematics in Science and Engineering.

[6] K. Masuda, *On the Global Existence and Asymptotic Behavior of Solutions of Reaction Diffusion Equations,* Hokkaido Math. J., 12 : 360-370 (1982).

[7] J. Morgan, *Global Existence for Semilinear Parabolic Systems,* SIAM J. Math. Anal., vol. 20, No. 5 : 1128-1144 (1989).

Department of Mathematics, B.P. 239
University of Nancy I
54506 - VANDOEUVRE-LES-NANCY
France
and also supported by
URA CNRS 750, Projet NUMATH, INRIA-Lorraine

Convexity of the Growth Bound of C_0-Semigroups of Operators

G. F. Webb

1. Introduction

If e^{tA}, $t \geq 0$ is a C_0-semigroup of bounded linear operators in a Banach space X with infinitesimal generator A, then the *growth bound* of e^{tA}, $t \geq 0$, defined as a function of A, is $\omega_0(A) := \lim_{t\to\infty} \frac{1}{t}\log(|e^{tA}|)$ (see [4], p. 619). If B is in $B(X)$ (the Banach algebra of bounded linear operators in X), then $A+B$ is the infinitesimal generator of a C_0-semigroup $e^{t(A+B)}$, $t \geq 0$ in X (see [7], p. 76). It is thus possible to consider $\omega_0(A+B)$ as a function of $B \in B(X)$ and investigate its properties. The question we investigate here concerns the following property of the growth bound: If $\alpha \in (0,1)$ and $B,\ C \in B(X)$, when is it true that

$$\omega_0(A+\alpha B+(1-\alpha)C) \leq \alpha\omega_0(A+B)+(1-\alpha)\omega_0(A+C). \quad (1.1)$$

If X is finite dimensional, then $\omega_0(A) = \sup\{Re\lambda : \lambda \in \sigma(A)\}$ (see [11], p. 171). In the finite dimensional case it is known that if A is an $n \times n$ nonnegative irreducible matrix and B and C are diagonal matrices, then (1.1) does hold (see [8], Corollary 1.1). That the convexity property (1.1) does not hold in general, even in the finite dimensional case, may be seen by the following example: Take $X = \mathbb{R}^2$, $\alpha = \frac{1}{2}$, and

$$A = \begin{bmatrix} 4 & 1 \\ 2 & 3 \end{bmatrix},\ B = \begin{bmatrix} 5 & 6 \\ 2 & 3 \end{bmatrix},\ C = \begin{bmatrix} 0 & 0 \\ 0 & 0 \end{bmatrix}.$$

Then $\omega_0(A+\frac{1}{2}B) = 9.106 > \frac{1}{2}\omega_0(A+B)+\frac{1}{2}\omega_0(A+C) = 6.5+2.5$.

G. R. Goldstein and J. A. Goldstein (eds.),
Semigroups of Linear and Nonlinear Operations and Applications, 259–270.

In [8] Kato proves the convexity property (1.1) under the following assumptions: $X = C(S)$, where S is a compact Hausdorf space, or $X = L^p(S)$, $1 \le p < \infty$, where S is a measure space, A is the infinitesimal generator of a C_0-semigroup of positive operators in X, and B and C are multiplication operators in X.

Our motivation in investigating the convexity property (1.1) arises from an age-structured model of a tumor cell population undergoing periodic chemotherapy treatment. In Section 2 we show that the convexity property (1.1) implies that in certain circumstances shorter periods of treatment are more effective than longer periods of treatment. In Section 3 we provide some simple sufficient conditions for (1.1) to hold. The general problem of determining conditions under which the inequality (1.1) holds, and when it is strict, remains an interesting open question.

2. A Model of Periodic Chemotherapy

Let $n(a,t)$ be the density with respect to age a of an age-structured tumor cell population at time t. The total population of tumor cells at time t is $\int_0^\infty n(a,t)da$. For the untreated tumor cell population the density $n(a,t)$ satisfies

$$n_t(a,t) + n_a(a,t) = -\beta(a)n(a,t), \tag{2.1}$$

$$n(0,t) = 2\int_0^\infty \beta(a)n(a,t)da, \; t > 0, \tag{2.2}$$

$$n(a,0) = \phi(a), \; a > 0, \tag{2.3}$$

where $\beta(a)$ is the rate of division of mother cells into two daughter cells and $\phi(a)$ is the initial age distribution of cells.

The problem (2.1)-(2.3) can be associated with a C_0-semigroup in $X := L^1(0,\infty)$. Let $\beta \in L^\infty_+(0,\infty)$ and let ess $\inf_{a>0}\beta(a) =:$

$\beta_0 > 0$. Define $A : X \longrightarrow X$ by

$$(A\phi)(a) = -\phi'(a) - \beta(a)\phi(a), \tag{2.4}$$
$$D(A) = \{\phi \in X : \phi' \in X \quad \text{and} \quad \phi(0) = 2\int_0^\infty \beta(a)\phi(a)da\}.$$

It is known that A is the infinitesimal generator of a C_0-semigroup $e^{tA}, t \geq 0$ in X and the formula $n(a,t) = (e^{tA}\phi)(a)$ provides a generalized solution to (2-1)-(2.3) (see [11], p 76). It is also known that e^{tA}, $t \geq 0$ has the following asymptotic behavior which cell biologists call asynchronous exponential growth (see [11], p. 188): Let λ be the unique real solution to the characteristic equation

$$2\int_0^\infty e^{-\lambda a}\beta(a)\exp[-\int_0^a \beta(b)db]da = 1. \tag{2.5}$$

Then $\lambda = \omega_0(A)$ and there exists a rank one projection P_A in X and constants $M \geq 1$ and $\epsilon > 0$ such that

$$e^{tA} = e^{t\omega_0(A)}P_A + e^{tA}(I - P_A),\ t \geq 0, \qquad \text{and} \tag{2.6}$$
$$|e^{tA}(I - P_A)| \leq Me^{t(\omega_0(A)-\epsilon)}, t \geq 0. \tag{2.7}$$

Now consider a periodic age-specific treatment of the tumor cell population, that is, a treatment which is periodic in time and which affects only cells in a certain age range. Such a treatment corresponds to the addition of a loss term to (2.1). Let $\alpha \in (0,1)$, let $\mu \in L^\infty(0,\infty)$, and for each $p > 0$ let

$$\mu(a,t;p) = \begin{cases} 0 & \text{if } a > 0 \text{ and } 0 \leq t \leq \alpha p \\ \mu(a) & \text{if } a > 0 \text{ and } \alpha p < t < p \end{cases} \tag{2.8}$$

Let $\mu(a,t;p)$ be defined for all $t \geq 0$ by periodicity with period p. The periodic treatment model (with period p) for this off-on type periodic treatment is

$$n_t(a,t;p) + n_a(a,t;p) = -(\beta(a) + \mu(a,t;p))n(a,t;p), \quad (2.9)$$

$$n(0,t;p) = 2\int_0^\infty \beta(a)n(a,t;p)da, \; t > 0, \quad (2.10)$$

$$n(a,0;p) = \phi(a), \; a > 0. \quad (2.11)$$

For each $p > 0$ define $(B(t;p)\phi)(a) = -\mu(a,t;p)\phi(a), \; \phi \in X$, and then (2.8)-(2.10) can be written abstractly as

$$\frac{d}{dt}U(t,s;p)\phi = (A + B(t;p))U(t,s;p)\phi, t \geq s \geq 0, \quad (2.12)$$

$$U(s,s;p)\phi = \phi \in X,$$

where $U(t,s;p), t \geq s \geq 0$ is an evolution operator in X. The question we consider here is whether treatment is more effective for shorter periods p or longer periods p. We will show that the convexity property (1.1) implies that treatment in the limiting case $p \longrightarrow 0$ is more effective than treatment in the limiting case $p \longrightarrow \infty$.

Define $C : X \longrightarrow X$ by

$$(C\phi)(a) = -\mu(a)\phi(a), \; \phi \in X \quad (2.13)$$

and observe that for $p > 0$

$$U(p,0;p)\phi = e^{(1-\alpha)p(A+C)}e^{\alpha pA}\phi, \; \phi \in X.$$

If $t = np$ for some integer n, then

$$U(t,0;p)\phi = (e^{t(1-\alpha)(A+C)/n}e^{t\alpha A/n})^n\phi, \phi \in X$$

and by the Trotter Product Formula (see Lemma 3.1 in Section 3) we obtain

$$\lim_{p=t/n \longrightarrow 0} U(t,0;p)\phi = e^{t(A+(1-\alpha)C)}\phi, \phi \in X. \quad (2.14)$$

Thus, for the limiting case $p = 0$ (which corresponds to continuous treatment) the value of $\omega_0(A + (1 - \alpha)C)$ determines the growth ($\omega_0(A + (1 - \alpha)C) > 0$) or extinction ($\omega_0(A + (1 - \alpha)C) < 0$) of the tumor.

For $p > 0$ we assume that the evolution operator $U(t, s; p), t \geq s \geq 0$ has the following asymptotic behavior: There exists $\lambda(p) \in \mathbb{R}$, $\psi_p \in X_+$, $\psi'_p \in X'_+$, and $\epsilon_p > 0$ such that $\|\psi_p\| = 1$, $U(p, 0; p)\psi_p = e^{\lambda(p)p}\psi_p$, and for all $\phi \in X$, as $t \longrightarrow \infty$

$$e^{-\lambda(p)t}U(t, 0; p)\phi = \langle\phi, \psi'_p\rangle\bar{u}(t; p) + O(e^{-\epsilon_p t}), \tag{2.15}$$

where $\bar{u}(t; p) = e^{-\lambda(p)t}U(t, 0; p)\psi_p$, $t \geq 0$ (see [3]). The growth or extinction of the tumor cell population undergoing treatment with treatment period p is thus determined by the Floquet constant $\lambda(p)$.

We claim that

$$\lim_{p \to \infty} \lambda(p) = \alpha\omega_0(A) + (1 - \alpha)\omega_0(A + C). \tag{2.16}$$

From (2.6) and (2.7) (which also hold for $e^{t(A+C)}$, $t \geq 0$) we have that for $p > 0$

$$\begin{aligned} e^{\lambda(p)p}\psi_p &= U(p, 0; p)\psi_p \\ &= e^{(1-\alpha)p(A+C)}e^{\alpha pA}\psi_p \\ &= \left[e^{(1-\alpha)p\omega_0(A+C)}P_{A+C} + e^{(1-\alpha)p(A+C)}(I - P_{A+C})\right] \\ &\quad \left[e^{\alpha p\omega_0(A)}P_A + e^{\alpha pA}(I - P_A)\right]\psi_p, \end{aligned} \tag{2.17}$$

and for some $\epsilon > 0$, as $p \longrightarrow \infty$

$$e^{[\lambda(p) - \alpha\omega_0(A) - (1-\alpha)\omega_0(A+C)]p}\psi_p = P_{A+C}P_A\psi_p + O(e^{-\epsilon p}). \tag{2.18}$$

From (2.18) we obtain

$$\limsup_{p\to\infty} \lambda(p) \le \alpha\omega_0(A) + (1-\alpha)\omega_0(A+C). \tag{2.19}$$

We next claim that

$$\liminf_{p\to\infty} \|P_{A+C}P_A\psi_p\| > 0. \tag{2.20}$$

From [13], p. 191, $P_A\phi = V(\phi)e^{-\lambda a}\pi(a,0)/M$, where $\lambda = \omega_0(A)$, $\pi(a,b) = \exp[-\int_b^a \beta(u)du]$, $V(\phi) = \int_0^\infty \int_b^\infty \beta(a)e^{\lambda(b-a)}\pi(a,b)\phi(b)$ $dadb$, and $M = \int_0^\infty a\beta(a)e^{-\lambda a}\pi(a,0)da$. Observe that for $\phi \in X_+$ and $\lambda > 0$

$$\begin{aligned} V(\phi) &= \|\phi\| - \lambda \int_0^\infty \int_b^\infty e^{\lambda(b-a)}\pi(a,b)\phi(b)dadb \\ &\ge \|\phi\| - \lambda \int_0^\infty \int_b^\infty e^{\lambda(b-a)}e^{\beta_0(b-a)}\phi(b)dadb \\ &= \frac{\beta_0}{\lambda+\beta_0}\|\phi\|. \end{aligned}$$

For $\phi \in X_+$ and $\lambda \le 0$, $V(\phi) \ge \|\phi\|$. Thus, there exists $K > 0$ such that $\|P_A\phi\| \ge K\|\phi\|$ for all $\phi \in X_+$. Let $P_A\phi = \langle\phi, \phi'_A\rangle\phi_A$ and $P_{A+C}\phi = \langle\phi, \phi'_{A+C}\rangle\phi_{A+C}$ for all $\phi \in X$, where ϕ_A, $\phi_{A+C} \in X_+$, $\|\phi_A\| = \|\phi_{A+C}\| = 1$, ϕ'_A, $\phi'_{A+C} \in X'_+$, and ϕ'_A, ϕ'_{A+C} are strictly positive on X_+ (see [6]). Then $\|P_{A+C}P_A\psi_p\| = \langle\psi_p, \phi'_A\rangle\langle\phi_A, \phi'_{A+C}\rangle = \|P_A\psi_p\|\langle\phi_A, \phi'_{A+C}\rangle \ge K\langle\phi_A, \phi'_{A+C}\rangle$, which implies (2.20). Then (2.18) and (2.20) imply

$$\alpha\omega_0(A) + (1-\alpha)\omega_0(A+C) \le \liminf_{p\to\infty} \lambda(p),$$

which together with (2.19) implies (2.16).

We next claim that

$$\omega_0(A + (1-\alpha)C) \le \alpha\omega_0(A) + (1-\alpha)\omega_0(A+C), \alpha\epsilon(0,1) \tag{2.21}$$

The convexity property (2.21) follows directly from Theorem 6.1 in [8]. We provide a direct proof here, however, in order to establish necessary and sufficient conditions for the inequality to be strict.

Theorem 2.1. Let $\beta, \mu \in L^\infty_+$, let ess $\inf_{a>0} \beta(a) > 0$, let A be defined as in (2.4), and let C be defined as in (2.13). Then (2.21) holds and the inequality is strict if and only if $\mu(a) \neq$ const.

Proof. Let $f(a) = \beta(a)\exp[-\int_0^a \beta(b)db]$, $a > 0$. From (2.5) we have that

$$1 = 2\int_0^\infty e^{-\omega_0(A)a} f(a)da, \tag{2.22}$$

$$1 = 2\int_0^\infty e^{-\omega_0(A+C)a} f(a)\exp[-\int_0^a \mu(b)db]da,$$

$$1 = 2\int_0^\infty e^{-\omega_0(A+(1-\alpha)C)} f(a)\exp[-(1-\alpha)\int_0^a \mu(b)db]da.$$

Let $x_0(a) = -\omega_0(A+(1-\alpha)C)a - (1-\alpha)\int_0^a \mu(b)db$, $x_1(a) = -\omega_0(A+C)a - \int_0^a \mu(b)db$, and $x_2(a) = -\omega_0(A)a$, $a \geq 0$. Since the exponential function is convex, for all $a \geq 0$

$$\exp((1-\alpha)x_1(a) + \alpha x_2(a)) \leq (1-\alpha)\exp(x_1(a)) + \alpha\exp(x_2(a)). \tag{2.23}$$

Then, (2.22) and (2.23) imply

$$2\int_0^\infty f(a)\exp((1-\alpha)x_1(a) + \alpha x_2(a))da \tag{2.24}$$

$$\leq (1-\alpha)\int_0^\infty 2f(a)\exp(x_1(a))da + \alpha\int_0^\infty 2f(a)\exp(x_2(a))da$$

$$= 1 = 2\int_0^\infty f(a)\exp(x_0(a))da.$$

Thus, (2.24) yields

$$\int_0^\infty f(a)\{\exp(x_0(a)) - \exp((1-\alpha)x_1(a) + \alpha x_2(a))\}da \geq 0, \tag{2.25}$$

which yields (2.21). If $\mu(a) \not\equiv$ const, then $x_1(a) \not\equiv x_2(a)$ for some $a_0 > 0$, and the inequality (2.23) is strict for a_0. Since $f(a) > 0$ for $a > 0$, (2.25) and (2.21) are also strict. If $\mu(a) \equiv$ const, then it can be shown easily that $\omega_0(A+(1-\alpha)C) = \alpha\omega_0(A)+(1-\alpha)\omega_0(A+C)$, since A and C, and hence e^{tA} and $e^{t(A+C)}$ commute. ∎

3. Sufficient Condition for Convexity of the Growth Bound

We suppose that A is the infinitesimal generator of a C_0-linear semigroup in X, B and C are bounded linear operators in X, and $\alpha \in (0,1)$.

Lemma 3.1. For each x in X and uniformly in bounded intervals of t

$$\lim_{n\to\infty} (e^{\frac{t}{n}\alpha(A+B)} e^{\frac{t}{n}(1-\alpha)(A+C)})^n x = e^{t(A+\alpha B+(1-\alpha)C)}x. \tag{3.1}$$

Proof. Define a new norm on X by

$$\|x\|_1 = \sup_{t\geq 0} e^{-t\omega_0(A)}\|e^{tA}x\|, \quad x \in X.$$

Then $\|\ \ \|$ and $\|\ \ \|_1$ are equivalent norms on X and $\|e^{tA}x\|_1 \leq e^{t\omega_0(A)}\|x\|_1$ for $x \in X$, $t \geq 0$. Let $A_1 := A - \omega_0(A)I$ and then $\|e^{tA_1}x\|_1 \leq \|x\|_1$ for $x \in X$, $t \geq 0$. From [7], Theorem 2.1, p. 495, $\|e^{t\alpha(A_1+B)}x\|_1 \leq e^{t\alpha|B|_1}\|x\|_1$, $\|e^{t(1-\alpha)(A_1+C)}x\|_1 \leq e^{t(1-\alpha)|C|_1}\|x\|_1$, and

$$\|e^{t(A_1+\alpha B+(1-\alpha)C)}x\|_1 \leq e^{t(\alpha|B|_1+(1-\alpha)|C|_1)}\|x\|_1$$

for $x \in X$, $t \geq 0$. Define $B_1 := B - |B|_1 I$ and $C_1 := C - |C|_1 I$. Then $\|e^{t\alpha(A_1+B_1)}x\|_1 \leq \|x\|_1$, $\|e^{t(1-\alpha)(A_1+C_1)}x\|_1 \leq \|x\|_1$, and

$\|e^{t(A_1+\alpha B_1+(1-\alpha)C_1)}x\|_1 \leq \|x\|_1$ for $x \in X$, $t \geq 0$. By the Trotter Product Formula (see [5], p. 53), for each $x \in X$ and uniformly in bounded intervals of t

$$\lim_{n\to\infty}(e^{\frac{t}{n}\alpha(A_1+B_1)}e^{\frac{t}{n}(1-\alpha)(A_1+C_1)})^n x = e^{t(A_1+\alpha B_1+(1-\alpha)C_1)}x,$$

which implies (3.1). ∎

Theorem 3.1. Suppose that for every $\epsilon > 0$ there exists $M_\epsilon \geq 1$ such for $t \geq 0$ and $n = 1, 2, \ldots$,

$$|(e^{t\alpha(A+B)/n}e^{t(1-\alpha)(A+C)/n})^n| \leq M_\epsilon e^{t(\alpha\omega_0(A+B)+(1-\alpha)\omega_0(A+C)+\epsilon)}. \tag{3.2}$$

Then (1.1) holds.

Proof. Let $\epsilon > 0$. By Lemma 3.1 and (3.2) there exists $M_\epsilon > 0$ such that for $x \in X$ and $t \geq 0$

$$\begin{aligned}
&\|e^{t(A+\alpha B+(1-\alpha)C)}x\| \\
&\quad = \lim_{n\to\infty}\|(e^{t\alpha(A+B)/n}e^{t(1-\alpha)(A+C)/n})^n x\| \\
&\quad \leq M_\epsilon e^{t(\alpha\omega_0(A+B)+(1-\alpha)\omega_0(A+C)+\epsilon)}\|x\|.
\end{aligned}$$

Since $\omega_0(A + \alpha B + (1-\alpha)C) = \inf\{\omega :$ there exists K_ω such that for $t \geq 0$, $|e^{t(A+\alpha\beta+(1-\alpha)C)}| \leq K_\omega e^{\omega t}\}$ (see [4], p. 619), (1.1) must hold. ∎

We remark that a sufficient condition for (3.2) to hold is that $e^{t(A+\alpha B)}$ and $e^{t(A+(1-\alpha)C)}$ commute for all $t \geq 0$. Another sufficient condition for (3.2) to hold is that $|e^{t(A+B)}| \leq e^{t\omega_0(A+B)}$ and $|e^{t(A+C)}| \leq e^{t\omega_0(A+C)}$ for all $t \geq 0$.

Theorem 3.2. Suppose that for $x \in X$, $t \geq 0$, and $s \geq 0$,

$$\|e^{s(A+B)}e^{t(A+C)}x\| \leq e^{t\omega_0(A+C)}\|e^{s(A+B)}x\|. \tag{3.3}$$

Then (1.1) holds.

Proof. Let $\epsilon > 0$ and define an equivalent norm $\| \ \|_\epsilon$ on X by $\|x\|_\epsilon = \sup_{s\geq 0} e^{-s(\omega_0(A+B)+\epsilon)}\|e^{s(A+B)}x\|$, $x \in X$. Then $\|e^{s(A+B)}x\|_\epsilon \leq e^{s(\omega_0(A+B)+\epsilon)}\|x\|_\epsilon$ for $x \in X$, $s \geq 0$. Further, (3.3) implies that $\|e^{t(A+C)}x\|_\epsilon \leq e^{t\omega_0(A+C)}\|x\|_\epsilon$ for $x \in X$, $t \geq 0$. Then, for $x \in X$, $t \geq 0$, $n = 1, 2, \ldots$,

$$\begin{aligned} &\|(e^{t\alpha(A+B)/n}e^{t(1-\alpha)(A+C)/n})^n x\|_\epsilon \\ &\leq e^{t(\alpha\omega_0(A+B)+(1-\alpha)(A+C)+\alpha\epsilon)}\|x\|_\epsilon. \end{aligned} \tag{3.4}$$

Since for every $\epsilon > 0$, $\|x\| \leq \|x\|_\epsilon$ and there exists M_ϵ such that $\|x\|_\epsilon \leq M_\epsilon\|x\|$, (3.4) implies (3.2), and hence (1.1). ∎

References

1. L. Cojocaru and Z. Agur, *A theoretical analysis of interval drug dosing for cell-cycle-phase-specific drugs*, to appear.

2. B. Dibrov, A. Zhabotinsky, Y. Neyfakh, M. Orlova, and L. Churikova, *Mathematical model of cancer chemotherapy. Periodic schedules of phase-specific cytotoxic-agent administration increasing to selectivity of therapy*, Math. Biosci. **73**(1985), 1-31.

3. O. Diekmann, H. Heijmans, and H. Thieme, *On the stability of the cell size distribution* II, Hyperbolic Partial Differential Equations III, Inter. Series in Modern Appl. Math. Computer Science, Vol. 12., M. Witten, ed., Pergamon Press (1986), 491-512.

4. N. Dunford and J. Schwartz, *Linear Operators, Part I: General Theory*, Interscience Publishers, New York, 1957.

5. J. A. Goldstein, *Semigroups of Linear Operators and Applications*, Oxford University Press, New York, 1985.

6. G. Greiner, *A typical Perron-Frobenius theorem with application to an age-dependent population equation*, Infinite-Dimensional Systems, Proceedings, Retzhof 1983, F. Kappel and W. Schappacher, eds., Lecture Notes in Mathematics, Vol. 1076, Springer-Verlag, Berlin Heidelberg New York Tokyo, 1984.

7. T. Kato, *Perturbation Theory for Linear Operations*, Springer-Verlag, Berlin Heidelberg New York, 1966.

8. T. Kato, *Superconvexity of the spectral radius, and convexity of the spectral bound and the type*, Math. Zeit. **180**(1982), 265-273.

9. R. Nussbaum, *Convexity and log convexity for the spectral radius*, Linear Algebra and Its Applications **73**(1986), 59-122.

10. A. Pazy, *Semigroups of Linear Operators and Applications to Partial Differential Equations*, Springer-Verlag, 1983.

11. G. Webb, *Theory of Nonlinear Age-Dependent Population Dynamics*, Monographs and Textbooks in Pure and Applied Mathematics Series, Vol. 89, Marcel Dekker, New York and Basel, 1985.

12. G. Webb, *An operator-theoretic formulation of asynchronous exponential growth*, Trans. Amer. Math. Soc. 303, No. 2 (1987), 751-763.

13. G. F. Webb, *Semigroup methods in population dynamics: Proliferating cell populations*, Semigroup Theory and Applications, Lecture Notes in Pure and Applied Math. Series, Vol. 116, Marcel Dekker, New York, 1989, 441-449.

Department of Mathematics
Vanderbilt University
Nashville TN 37240 U.S.A.

Index

A-B-L-P Theorem 233,235,237

Absorbing set 164

Abstract Cauchy problem 9, 59, 217,220,221

Abstract d'Alembert formula 48

Abstract evolution equation 164

Abstract inhomogeneous Cauchy problem 74

Abstract wave equation 46

Action 28

d'Alembert's formula 49

Age-structured model 260

α-times integrated semigroup 205,218-221

Analytic semigroup 14

Antimaximum principle 115

Approximation Theorem 20, 79-81

Arendt-Batty 235

Asymptotic equipartition of energy 48

Asymptotically $\infty - r$ stable 147

Asynchronous exponential growth 251

Backward difference scheme 62

Balance law 251

Banach (see Picard)

Barenblatt solution 95

G. R. Goldstein and J. A. Goldstein (eds.),
Semigroups of Linear and Nonlinear Operations and Applications, 271–283.

Bénilan 63

Bessel's equation 213

Bessel-type differential operator 208,213

Bifurcation 99,133

Bound state 40

Boundary conditions

inhomogeneous periodic 88

nonlinear 34,88

Boundary group 206,207

Boundary value 205,206

Boundary value problem 111

Branch of solutions 113

C-regularized 121

(C_o)-semigroup 10,61

Cell population 260

Center manifold 195

Center unstable manifold 195

Central Limit Thoerem 25

Chernoff formula 22

Conservation law 65

Conservation of mass 93

Continuation 241

Contraction semigroup 11,61

Convexity 259

Crandall-Liggett Theorem 63

Crandall-Tartar Theorem 94

Critical points 110

D'Alembert's formula 49
Daletskii (see Lie)
Degenerate parabolic problem 83
Degree 115
Difference scheme
  backward 62
  forward 62
Dirichlet Laplacian 32
Dirichlet problem 109
Discrete spectrum 231
Dissipative 11,63
Dissipative chemical reaction 152
Distribution function 24
Dual integral equations 216
Duality map 68
Duhamel formula 36

Elliptic semilinear problem 100
Ellipticity 18
Energy 16,46
Energy norm 17,60
Entropy condition 66
Equipartition of energy 48
Erdélyi-Kober operators 216
Evolution operator 262,263
Evolution problem 203
Exponentially asymptotically stable 187
Exponentially bounded C-regularized semigroup 121

Favard class 78
Fermi pseudopotentials 99
Feynman integral 28
Feynman path formula 27
Finite speed of propagation 61,95
Floquet constant 263
Flow through porous medium 60
Forward difference scheme 62
Forward invariant region 144
Fractional derivative 208
Fractional integral 205,207,208,216,221
Fractional power 205,220,221
Fractional power semigroup 205,214

Gaussian heat kernal 60
Gelfand 234
Generalized domain 78
Generator 10,67,215,217-220,241
  C-regularized semigroup 121
  infinitesimal 10,67,133
Glicksberg 221
Global attractor 162
Global dynamics 162
Global existence 251
Green's function 86
Gronwall-type inequality 77

Group
  boundary 206,207
  unitary 14,67
Growth bound 229,259

Hamilton-Jacobi equation 60
Harvesting 109
Heat equation 59
  infinite dimensional 129
Heat semigroup 132
Hilbert-Schmidt operator 38
Hille 62
Hodgkin-Huxley equations 162
Hodgkin-Huxley system 160,163
Holomorphic 241
Holomorphic C_0-semigroup 177
Holomorphic semigroup 205-207,215
Hopf equation 66
H-smooth 44
Hyper-Bessel operator 213

Index law 206,211
Infinite dimensional heat equation 129
Infinite speed of propagation 60,95
Infinitesimal generator 10,67,133
$\infty - r$ stable 146
Inhomogeneous periodic boundary conditions 88
Integral solution 76

Integrated semigroup 54
Interpolation spaces 187
Invariant manifold 198
Invariant region
  forward 144
Irreversible part 231,233

Jacob-Glicksberg-deLeeuw splitting 230

Kato-Kuroda Theorem 45
Katznelson-Tzafriri 234
Kinetic energy 46,60
Kobayashi 63
Kronecker set 232

Laplace operator 204
Laplace transform 218
Laplacian 109
Lebesgue (see Riemann)
Lebesgue decomposition 41
Leeuw, de 230
Leray-Schauder degree 115
Liapunov Stability Theorem 226,228
Lie-Trotter-Daletskii product formula 27
Liouville (see Riemann)

Lipschitz seminorm 63
Lipschitzian 109
Lotka-Volterra system 153
Lyapunov functional 148,149
Lyapunov-Perron approach 186
Lyubich-Phong 235

M-dissipative 11,63
Marcinkiewicz space 92
McArthur (see Rozenwig)
Mean Ergodic Theorem 37
Mellir convolution 209
Mellir transform 208,214
Mild solution 30,76
Multiplication operator 13
Multiplier 209,210,212,217
Multivalued function 65,68

Navier-Stokes equation 31,88
Nikodym (see Radon)
Nodal hypersurface 114
Nonautonomous problem 136
Nonlinear boundary condition 34,88
Nonlinear conservation law 65
Nonlinear eigenvalue problem 99
Nonlinear evolution problem 203
Nonlinear parabolic equations 185

Nonlinear semigroups 59
Nonlinearity 109
Nonnegative solution 110
Normalized duality map 68

Parabolic equation 185
Partially dissipative system 165
Periodic chemotherapy 260
Peripheral spectrum 234,237
Perron (see Lyapunov)
Perturbation Theorem 17,79,81-83
Phong (see Lyubich)
Picard-Banach Fixed Point Theorem 82
Point interactions 99
Porous medium equation 60,62
Positive semigroup 42
Positive solution 112
Positivity preserving semigroup 42,94
Positone 109
Potential energy 46,60
Principle of linearized stability 187
Pseudo-differential operator 209

Quasianalytic collection 244
Quasiholomorphic 241,242

Radially symmetric 112
Radon-Nikodym property 63
Reaction-diffusion system 143,251
Regular semigroup 207
Resolvent identity 71
Reversible part 231,233
Riemann-Lebesgue Lemma 207
Riemann-Lebesgue operator 48
Riemann-Liouville fractional integral 205,207,208,216,221
Rozenwig-McArthur 109

Scattered state 40
Scattering theory 43
Schauder (see Leray)
Schauder Fixed Point Theorem 86
Sectorial 186
Self-adjoint 15
Semigroup 10,241
  α-times integrated 205,218-221
  analytic 14
  C-regularized 121
  C_0 10,61
  contraction 11,61
  exponentially bounded C-regularized 121
  fractional power 205,214
  holomorphic 241
  holomorphic C_0 177

Semigroup (continued)
integrated 54
nonexpansive 245
nonlinear 59
positive 42
positivity preserving 42,94
regular 207
strongly continuous 61
type ω 71,73
ultrapower 236
Semilinear problem 100
Semipositone 109
Separable function 150
Similarity 13
Singular solutions 100
Singularly pertubed Hodgkin-Huxley equations 163
Smooth bounded region 115
Smoothing 138,139
Solution
Barenblatt 95
branch of 113
integral 76
mild 30
nonnegative 110
positive 112
singular 110
stable positive 112
subsolution 109
sub-super 109
unstable positive 112
weak 66

Space
  interpolation 187
  weak L^p 92
  weighted 207,208,214
Spatially degenerate parabolic problem 83
Spectral bound 226,230
Spectral Theorem 16
Speed of propagation
  finite 61,95
  infinite 60,95
Stable 228,236
  asymptotically $\infty - r$ 147
  exponentially asymptotically 187
  $\infty - r$ 146
  manifold 187
  strongly 234
  super 236
Stable positive solution 112
Stokes operator 32
Stone's Theorem 16
Strong solution 74
Strongly continuous semigroup 61
Strongly elliptic 177
Strongly mixing 232
Strongly stable 234
Subcritical 112
Subsolution 109
Sub-super solutions 109
Superlinear 110
Superreflexive 236
Superstable 236

Symmetric 15
Symmetry breaking 109
Systems of reaction-diffusion equations 143

Telegraph equation 51
Trajectory 242
Trotter (see Lie)
Trotter product formula 262,267
Trunkation 145
Tzafriri (see Katznelson)

Unitary group 14,67
Unstable 187
Unstable manifold 187
Unstable positive solution 112

Variation of constants formula 186
Variation of parameters formula 29
Volterra (see Lotka)

Wave equation 59
Wave function 27
Weak L^p-space 92

Weak 227

Weak Mixing Theorem 228

Weak solution 66

Weak Spectral Mapping Theorem 234

Weakly mixing 232

Weighted spaces 187,207,208,214

Well-posed 10,59

Wiener's Theorem 39

Yosida (see Hille)

Yosida approximation 77,90

MIX
Papier aus verantwortungsvollen Quellen
Paper from responsible sources
FSC® C105338

If you have any concerns about our products,
you can contact us on
ProductSafety@springernature.com

In case Publisher is established outside the EU,
the EU authorized representative is:
Springer Nature Customer Service Center GmbH
Europaplatz 3, 69115 Heidelberg, Germany

Printed by Libri Plureos GmbH
in Hamburg, Germany